FOOD AND BEVERAGE CONSUMPTION AND HEALTH

CITRUS FRUITS

PRODUCTION, CONSUMPTION AND HEALTH BENEFITS

FOOD AND BEVERAGE CONSUMPTION AND HEALTH

FOOD AND BEVERAGE CONSUMPTION AND HEALTH

CITRUS FRUITS

PRODUCTION, CONSUMPTION AND HEALTH BENEFITS

DAPHNE SIMMONS
EDITOR

nova publishers
New York

Library of Congress Cataloging-in-Publication Data

Library of Congress Control Number: 2015955200

ISBN: 978-1-63484-078-1

Published by Nova Science Publishers, Inc. † New York

CONTENTS

Preface

Citrus is the most widely produced fruit in the world and it is grown in more than 80 countries. Due to its varied and wide chemical composition as a consequence of its nature, citrus is an exceptional feedstock to the designing and assessing of biorefineries. A wide spectrum of products are obtained from citrus, which nowadays are extracted and purified into essential oils, antioxidants and other compounds. This book provides research on the production, consumption and health benefits of citrus fruits. The first chapter begins with an overview of citrus based refineries. Chapters two and three discuss hesperidin and narirutin, which are citrus flavonoids. Chapter four studies the use of citrus residues as raw materials for biomolecules and energy. Chapter five collects information from published works about the alternative use of citrus residues as efficient and promising adsorbents in clean water technology. The final chapter examines citrus genetic improvement.

Chapter 1 - Citrus is the most widely produced fruit in the world and it is cultivated in more than 80 countries. Brazil leads in citrus production, with more than 18.90 million metric tons of fruit produced during 2004–05, followed by the United States and China. Brazilian citrus production is oriented toward processing, while USA citrus production is focused toward processing and the fresh fruit market. Nowadays Colombia is a smallholder producer compared to Brazil and USA, nevertheless many expansion possibilities appear in the west zones of the country. Citrus production in Colombia was around 187 million tons for 2010. Nowadays citrus agroindustry in Colombia is not a well-established chain and many opportunities appear. On the other hand, citrus is one of the most exceptional feedstock to design and assess biorefineries due to its varied and wide chemical composition as a consequence of its nature.

From citrus are obtained a wide spectrum of products, which nowadays are extracted and purified such as essential oils, antioxidants and other value-added compounds as pectin. It is also important to obtain products for human consumption to guarantee food security, such as concentrated juices factories which has the major producers in Brazil and USA. Therefore, the aim of this chapter is to evaluate a citrus-based biorefinery for the integrated production of essential oil, concentrated juice, antioxidant, citrus seed oil, pectin, xylitol, PHB, ethanol, citric acid, lactic acid and electricity. The evaluation consists in the influence of energy and mass integration on the economical feasibility, environmental impact and possible social aspects that contribute in some way in rural development and food security preservation.

Chapter 2 - Hesperidin is the principal bioflavonoid found in citrus fruits, with very interesting bioactivity properties that still are the object of intensive research. Hesperidin and its aglycon form, hesperetin, are present in large quantities in oranges (*Citrus sinensis*) in particular. In young, immature oranges, these flavones can account for up to 14% of the fruit weight. Hesperetin is the 3',5,7-trihydroxy-4'-methoxy flavanone, and hesperidin contains not only the flavanone moiety, but also a rutinose disaccharide that has one D-rhamnose united with glycoside bond to the D-glucose unit. This paper presents and discusses chemical and physical properties of the orange bioflavones, as well as the most common methods of isolating and purifying these compounds. As a secondary plant metabolite, hesperidin is produced as a protective agent in citrus, and its defense role and biosynthesis will also be briefly discussed here. Many interesting bioactive properties of these phytochemicals have been reported, including antioxidant, anti-inflammatory, hypolipidemic, vasoprotective and anticarcinogenic properties, and an extensive review of these properties will be presented. Last but not least, the authors will present the most up-to-date developments in the research field that account for the mechanisms of action of these compounds.

Chapter 3 - *Citrus unshiu* is one of the most important varieties of citrus grown in Northeast Asia. Its peel is known as 'Chinpi,' a non-toxic edible ethnopharmaceutical herb in China and Korea, and has been clinically used as a traditional medicine to treat common cold, dyspepsia, cough and phlegm. Modern therapeutic studies have proven that citrus flavonoids have anti-oxidative, anti-inflammatory and anti-allergic activities. In this chapter, an efficient way to isolate citrus flavonoids, narirutin and hesperidin, from *Citrus unshiu* was introduced. Physiological properties such as anti-inflammatory activities and anti-alcoholic liver disease were also reviewed with suggestions on improving their bioavailability in a body through enzymatic modifications.

Chapter 4 - The replacement of the fossil-based raw materials either fully or partially is an objective in many countries, being of special interest the use of local biomass such as agricultural, forest, agro-industrial and industrial wastes, due to its low cost and large availability. According to FAOSTAT, by 2011 approximately 120 million tons of citrus were produced worldwide, with oranges accounting approximately 63.1 million tons. Approximately 60% of the total citrus production is market for fresh consumption, while the other 40% is used in the agroindustry to extract no more than the 50% of the fruit weight as juice. Residues from agroindustrial processing are composed by peel, seeds and remaining pulp and, in most of the cases, are used to spread soils, to produce animal feed, or to be burned. However, these conventional disposal methods can cause negative effects on the soil and superficial waters. Moreover, several value-added products, such as phytochemicals, pharmaceuticals, food products, essential oils, seed oil, pectin and dietary fibers, can be obtained from orange residues. In this chapter, simulation results of the production of biofertilizers, gibberellic acid and electricity from orange peel as stand-alone products are presented. Moreover, the experimental characterization was assessed. Results from the characterization procedures have been used to feed the simulations to obtain the mass and energy balances that were subsequently used to perform the economic and environmental analysis of the above mentioned processes. Moreover, comparisons from the techno-economic and environmental points of view of the stand-alone processes were performed. Besides, and based on the experimental results of the physicochemical characterization, two biorefinery schemes were techno-economic and environmentally evaluated.

Chapter 5 - Water pollution is still a serious problem for the entire world. Adsorption technology is a promising process which based on fabrication of novel, cheap, non dangerous and highly sorptive materials for application in wastewater purification processes. *Citrus* species generally produced for the fresh consumption or the production of fruit juice but also have lot of application in medicine, food processing and agriculture sectors. This review collects information from published works about the alternative use of *Citrus* residues as efficient and promising adsorbents in clean water technology.

For this purpose, isotherm (Langmuir, Freundlich, etc.), kinetic (pseudo-first, -second order, etc.), thermodynamic (free energy Gibbs, enthalpy, entropy) and desorption-regeneration studies were discussed in detailed. Moreover, significant factors such as pH, agitation time, temperature, adsorbent dosage and initial dye concentration are also reported extensively.

Chapter 6 - The Citrus genetic improvement is obtained throughout the application of several breeding procedures of extant species. Main aims of such breeding approaches are to obtain seedless fruits with easily removable peel, optimal size, excellent and original organoleptic characters, and possibly fruits endowed with precocious or late ripening. Citrus fruits and some of their transformation products, such as juices, fall in the large category of the functional foods owing to their content of important secondary metabolites defined nutraceutical components, whose beneficial effects on the human health are continuously evidenced. In this context the aim of the breeding processes is to obtain new varieties with an increased amount of nutraceutical components. Besides these characters mainly associated to the new fruits, other important agronomic and economic aspects concern the production of plants with high productivity and improved resistance against biotic and abiotic stresses.

On these bases, the authors' groups have focused the research activity in the genetic improvements of high quality cultivars and the production of new citrus fruits, namely hybrids. In particular, the authors' interest, has been addressed to the study of the chemical composition (mainly polyphenols from juices and peel essential oils) of new Citrus hybrids, with the aim of an exhaustive phytochemical characterization and, possibly, the evaluation of these new fruits for their introduction into the fresh market and into the industrial chain of transformation.

The new hybrids have been obtained through somatic hybridization by protoplast fusion. This technique, enabling to combine fully or partially, nuclear and cytoplasmic genomes at the interspecific and intergeneric levels, allows to widen the gene pool and to increase the genetic diversity of a species, circumventing the naturally occurring sexual incompatibility barriers (nucellar polyembryony, long juvenility and pollen/ovule sterility). Following this approach, the authors' breeding program has given rise to dozens of somatic hybrid and cybrids that are now being evaluated for their agronomic and productive characters.

A wide description of the different adopted breeding strategies and a summary of the phytochemical analyses of the new varieties obtained in these last years will be given.

In: Citrus Fruits
Editor: Daphne Simmons

ISBN: 978-1-63484-078-1

Chapter 1

CITRUS BASED BIOREFINERIES

Jonathan Moncada, Valentina Hernández, Yessica Chacón, Ramiro Betancourt and Carlos A. Cardona*
Instituto de Biotecnología y Agroindustria,
Universidad Nacional de Colombia Sede Manizales,
Manizales, Colombia

ABSTRACT

Citrus is the most widely produced fruit in the world and it is cultivated in more than 80 countries [1]. Brazil leads in citrus production, with more than 18.90 million metric tons of fruit produced during 2004–05, followed by the United States and China. Brazilian citrus production is oriented toward processing, while USA citrus production is focused toward processing and the fresh fruit market [1]. Nowadays Colombia is a smallholder producer compared to Brazil and USA, nevertheless many expansion possibilities appear in the west zones of the country. Citrus production in Colombia was around 187 million tons for 2010 [2]. Nowadays citrus agroindustry in Colombia is not a well-established chain and many opportunities appear. On the other hand, citrus is one of the most exceptional feedstock to design and assess biorefineries due to its varied and wide chemical composition as a consequence of its nature.

From citrus are obtained a wide spectrum of products, which nowadays are extracted and purified such as essential oils, antioxidants and other value-added compounds as pectin. It is also important to obtain products for human consumption to guarantee food security, such as concentrated juices factories which has the major producers in Brazil and USA. Therefore, the aim of this chapter is to evaluate a citrus-based biorefinery for the integrated production of essential oil, concentrated juice, antioxidant, citrus seed oil, pectin, xylitol, PHB, ethanol, citric acid, lactic acid and electricity. The evaluation consists in the influence of energy and mass integration on the economical feasibility, environmental impact and possible social aspects that contribute in some way in rural development and food security preservation.

* Corresponding author: Tel: +57 6 8879400x55880; E.mail address: ccardonaal@unal.edu.co (Carlos A. Cardona).

Keywords: mandarin, biorefinery, value-added products

1. INTRODUCTION

1.1. The Biorefinery Concept

Depending on the physical and chemical nature of the raw material as well as on the economic interest, its yields and distributions vary widely. However, the term biorefinery could be extended to other sectors at the industrial scale, if products that only can be obtained from vegetable raw materials and foodstuffs are included [20, 21]. Sustainable multiproduct biorefineries should focus on large portions of biomass that will produce multiple streams with large volumes and lower market prices (e.g., biofuels) and streams with low volumes and high market prices (e.g., biomolecules) [22-24].

Huang et al. [25] defined biorefinery as processes that use bio-based resources such as agriculture or forest biomass to produce energy and a wide variety of precursor chemicals and bio-based materials, similar to the modern petroleum refineries. Industrial platform chemicals such as acetic acid, liquid fuels such as bioethanol and biodegradable plastics such as polyhydroxyalkanoates can be produced from wood and other lignocellulosic biomass. In compliance with Huang and González-Delgado & Kafarov [25, 26] a biorefinery is the most promising way to create a biomass-based industry. Other authors [11, 15, 17, 18, 21, 24, 26, 27, 29, 30, 32-45] conceive a biorefinery as a facility that integrates biomass conversion processes and equipment to produce fuels, power, and value-added chemicals from biomass. For this point of view, the biorefinery concept is analogous to crude oil refineries, which produce multiple fuels and products from petroleum. In a broad definition biorefineries process all kinds of biomass (all organic residues, energy crops, and aquatic biomass) into numerous products (fuels, chemicals, power and heat, materials, and food and feed). A biorefinery is a conceptual model for future biofuel production where both fuels and high-value coproduct materials are produced. Biorefineries would simultaneously produce biofuels as well as bio-based chemicals, heat, and power. Officially, the US Department of Energy (DOE) uses the following definition: *"A biorefinery is an overall concept of a processing plant where biomass feedstocks are converted and extracted into a spectrum of valuable products based on the petrochemical refinery."* Besides, The American National Renewable Energy Laboratory (NREL) published the definition: *"A biorefinery is a facility that integrates biomass conversion processes and equipment to produce fuels, power, and chemicals from biomass. The biorefinery concept is analogous to today's petroleum refineries, which produce multiple fuels and products from petroleum. Industrial biorefineries have been identified as the most promising route to the creation of a new domestic biobased industry"* [41].

The Biorefinery is a complex system where biomass is processed to obtain energy, biofuels and value-added products. This concept can be compared to the current concept of oil refineries where the processes are based on the fractioning of a complex mixture. However, there are two major elements that make them different: firstly the raw material, because those used in biorefinery have not undergone the biodegradation of crude oil over the time. So the possibilities of obtaining more products using biomass as a feedstock are greater;

and the second is the application of different existing and emerging technologies in order to obtain bioproducts.

Furthermore biorefining involves assessing and using a wide range of technologies to separate biomass into its principal constituents (carbohydrates, protein, triglycerides, etc.), which can subsequently be transformed into value-added products. The palette of products from a biorefinery not only includes the products obtained in an oil refinery, but also products that cannot be obtained from crude. Biorefineries can produce energy in the form of heat or by producing biofuels, molecules for fine chemistry, cosmetics or medicinal applications, materials such as plastics and sources of human food and animal feed [27-32].

Biorefineries would present more economical options, where bio-based chemicals are co-products of liquid fuel. Future biorefineries would be able to mimic the energy efficiency of modern oil refining through extensive heat integration and co-product development. Heat that is released from some processes within the biorefinery could be used to meet the heat requirements for other processes in the system.

However, the definition of the term biorefinery has been a subjected to debate. Ideally, a biorefinery should integrate biomass conversion processes to produce a range of fuels, power, materials, and chemicals from biomass. Conceptually, a biorefinery would apply hybrid technologies from different fields including polymer chemistry, bioengineering and agriculture. Simply, many petrochemicals are produced from crude oil-fed refineries, whereas in the future, it is anticipated that many bio-based products analogous to petrochemical will be produced in biorefineries fed with biomass.

The term biorefinery is derived from both the feedstock which is renewable biomass and also the bioconversion processes often applied in the treatment and processing of the raw materials. This allows the development of systems that ideally attempt to render the term ‘‘waste”, in its application to biomass processing, obsolete as each production stream has the potential to be converted into a by-product stream rather than waste streams [5, 15, 18, 21, 26, 27, 29, 30, 32, 33, 35-46].

Generally, a biorefinery approach involves multi-step processes in which the first step, following feedstock selection, typically involves treating the precursor-containing biomass to make it more amenable for further processing. This step is conventionally referred as pretreatment. Following pretreatment, the biomass components are subject to a combination of biological and/or chemical treatments. The outputs from this step (specialty chemicals or reducing sugars) could be further converted into chemical building blocks for further processing uses. Additionally, the conversions to specialty polymers ready for market use, to a fuel/energy source, or use in composite materials are possible processing options [21, 47-52].

By integrating production of value-added bioproducts into biorefineries with fuel and power output, overall profitability and productivity of all energy related products are potentially improved. Increased productivity and efficiency can also be achieved through operations that decrease overall energy intensity of biorefineries unit operations, maximizing use of all feedstock components, byproducts and waste streams, and using scale-up economies, common processing operations, materials, and equipment to drive down all production costs. Biorefinery can be considered as an evolution of concepts like “*Green Chemistry*” or *Chemurgy* [33, 52-56].

Table 1. Feedstock classification into the different generations

First generation: Edible Crops	Second generation: Residues (Mainly lignocelluloses)	Second generation: Non edibles	Third generation: Algae	Ref.
Corn stover, Oil Palm, Sugar beet, Sugarcane, Sorghum	Sawdust. Starchy residues. Woody biomass. Crop residues	Jatropha Castorbean	-	[55]
Wheat, straw, switchgrass	Fibers. Pulping liquors	-	-	[43]
Soybean. Sunflower. Rapeseed	-	-	-	[31]
-	Forest harvesting residues Bagasse Paper	-	-	[44]
Alfalfa		Soapnut Soap stock Karanja		[60]
-	-	-	*Botryococcus braunii. Crypthecodinium Nitzschia sp. Phaeodactylum Schizochytrium sp. Tetraselmis suecia Pavlovalutheri*	[26]
-	-	-	*Scenedesmus ob. Spirulina maxima. Ankistrodesmus sp. Chaetoceros cal. Chlorella vulgaris. Dunaliella tertiolecta.*	[57]
Animal fats	Waste cooking oils Fats from slaughterhouse	Grease traps		[58]
-	Raw Glycerol from biodiesel	-		[6]
	Urban wood wastes Industrial organic wastes			[59]
Sugarcane	Sugarcane biomass. Cell biomass from fermentations	-	-	[11]

1.2. Feedstocks and Products

A biorefinery must follow a holistic approach including new challenges to account for the wide range of raw materials and the need to develop patterns of local and regional solutions. These biorefineries will likely take the form of the design of regional development to better exploitation of resources. Therefore the first level that must be evaluated is the feedstock. Feedstocks can be classified in three types. The first type of feedstocks refers to crops, determined as the first generation. The first generation feedstocks also make reference to crops which are destined to food processing to preserve food security. The second type of feedstocks (so called second generation feedstocks) makes reference to agro-industrial residues from the harvesting and processing of first generation materials, for instance lignocellulosic biomass. Also the second type of feedstocks makes reference to crops that do not need special treatment and do not threat with food security, as the case of some oilseeds (e.g., Jatropha Curcas, Castorbean). The third and last type considered for this approach involves the uses of algae for several metabolites production, referred as the third generation feedstocks. A multiproduct biorefinery from algae can be raised because the same species of algae are capable to synthesize multiple varieties of products. Additionally, the residues generated in the algae processing can be integrated with second generation feedstocks [6, 11]. Examples of feedstock classification are summarized in Table 1.

Table 2. Product classification into shown in the literature for biorefinery examples

Biofuels	Food products	Bioenergy	Biomaterials	Biochemicals	Ref.
Biodiesel, Bioethanol	Gluten	Electricity, Heat	Activated carbon	Oxy fuel additives, Phenols and furfural, Fatty acids, Industrial surfactants. Dyes and pigments. Oils and inks. Paints and varnishes. Detergents-Cleaners	[61]
Syngas	Proteins Aminoacids Sugar substitutes	-	-	-	[40]
Lubricants Bioethanol Biomethanol	-	-	-	-	[62]
Hydrogen	-	-	-	-	[42]
Glycerol	-	Steam Power	Pulp and papers	Agricultural chemicals, Fertilizers, Sorbitol	[32]
Glycerol, Biomethane, Lignin	-	Charcoal	Industrial adhesives Biopolymers films, Composite materials, Dielectric fluids, Binders	Pharmaceuticals, Cosmetics	[54]
-	-		Bioplastics, Biobased epoxy, resin, Cement	-	[63]
Vanillin	Protective colloids Thickeners Emulsifiers Stabilizers				[50]
Bioethanol	Sugar	Electricity Steam	Poly-3-hydroxybutirate	Anthocyanins	[19]

The analysis of feedstocks considers possible relations between the different generations (first, second and third generations). This establishes different sequences to obtain different products based on the affluence of diverse material flows. For this study, six families of products are considered: biofuels, bioenergy (referred as direct energy), biomolecules and natural chemicals, biofertilizers, biomaterials and food products. Table 2 shows the classification of products for different examples of biorefineries. From the direct relationship between feedstocks and products, biorefineries can be also classified: first generation, second generation and third generation.

1.3. Technologies

Depending on raw materials, technological processes, and products obtained, biorefinery platforms can be distinguished based on sugar (biochemical), syngas (thermochemical), biogas, or carbon-rich chains platforms. Biorefinery platforms may incorporate other processes from other platforms and combined different processing routes. Some biorefinery platforms are described as lignocellulose feedstock based biorefinery, whole crop biorefinery that uses cereals integrating residues and as an alternative feedstocks generation among many others [53].

2. Methodology Description

2.1. Process Design Approach

In this chapter, twelve scenarios were assessed (Six scenarios for orange and other six for mandarin as feedstocks). Scenarios follow the technological description on distribution shown in the following sections. The evaluation of the scenarios consist on the impact of energy integration, mass integration and energy integration plus cogeneration systems. Each technological scheme was evaluated from the techno-economic and environmental points of view. For all scenarios feedstock consists in 100 tons/h of fresh citrus fruit (for orange and mandarin). Feedstock quantity is very large in proportion to the current Colombian plantations, representing approximately the 430% of Colombian productivity, which is very low and is not competitive in the World market. Nevertheless, this is an interesting opportunity to show that potential citrus crops can be expanded and job generation through crop plantations leads to an interesting social benefit. On the other hand, for well-established citrus agroindustry chains in countries such as Brazil (Sao Paulo) and USA (Florida), the biorefinery configuration described may help in integral residue uses and different processing sequences in contrast to the current ones. Therefore, feedstock flowrate represents the 4.42 and 10.69% of the total production in Brazil and USA, respectively. Then, the results showed in this chapter can be extended to Brazil and USA as the most important worldwide referents.

Table 3. Scenarios generated through the combination of energy and mass integration possibilities

Energy Integration Level			Mass integration Level		Feedstock	
No integration	Full integration	Full integration + cogeneration	No integration	Full integration	Orange	Mandarin
X			X		Sc-1	Sc-7
	X		X		Sc-2	Sc-8
		X	X		Sc-3	Sc-9
X				X	Sc-4	Sc-10
	X			X	Sc-5	Sc-11
		X		X	Sc-6	Sc-12

Considering the above mentioned, scenario description considers three levels of energy integration, and two levels of mass integration. The combination of these levels makes the six scenarios per feedstock (orange and mandarin) obtaining 12 scenarios. The energy integration levels consist in a first level were no energy integration is considered, a second level includes full energy integration based on the composite curves which relates hot and cold stream in the process, but in this level no cogeneration system is coupled with the biorefinery system. A third level consists in full energy integration plus cogeneration from solid residues as lignin and cell biomass from fermentation processes. Last is based on process flowsheet where cell biomass from fermentations is used as feedstocks in cogeneration plants as shown by Moncada, et al. [37]. Mass integration levels consist in a first level where no mass integration is done for water. The second level includes a water treatment section where water from different processing plants are recovered and recycled. To a better understanding of scenario description Table 3 shows the combination of the different levels to build up the scenarios for both feedstocks.

2.2. Process Description

In this chapter a proposal of a technological sequence for obtaining 11 value-added products from orange and mandarin are shown: concentrated juice, essential oil, antioxidants, seed oil and pectin as products extracted from the dry fruit. Therefore, the last products are integrated with products based in a platform of sugars obtained from hydrolysis of the solid wastes: xylitol, ethanol, PHB, citric acid and lactic acid.

The solid wastes are evaluated in the integration of the cogeneration process for electricity and heat production to be used in the plant. In order to compare orange and mandarin as feedstock, Table 4 shows the average chemical composition of each friut and Table 5 shows the percentage of seed, pulp and peel for both of them.

To understand the selected distribution of the process, this sequence in the transformation stages is explained as follow. The first step is the reception of the feedstock in which the entire fruit is received. After it is carried out a pulping process to separate the seeds, pulp and peel. Once these three fractions are obtained, the pulp is used in the concentrated juice production plant. Resulting streams consist in fiber and concentrated juice. The juice is commercialized while the fiber is used as raw material for the pectin production plant. On the other hand, the peel from pulping is sent to the plant for essential oil extraction to extract the volatile fraction present in the peel. Until now, it is evident that this sequence preserves the characteristics and the importance of these products. Moreover the applications of these products in food and pharmaceutical industries require high grade purity. On the other hand, the solid material remaining of the extraction of essential oil is rich in flavonoids and antioxidants.

Therefore, the solids remaining from the peel are mixed with the solid material from seeds for obtaining antioxidants and oil. The solid material resulting from the last processes are still rich in polysaccharides such as pectin and lignocellulosic complex. Therefore, these characteristics are exploited in the pectin extraction process. Once the pectin is extracted, a solid material, a rich polysaccharides liquor and soluble sugars are obtained. This stream is treated in the sugar extraction plant using acid hydrolysis to produce xylose and glucose as main products. From here, five products are derived based on the platform of pentoses and hexoses. The xylose obtained in the acid hydrolysis is sent to the xylitol production process.

Table 4. Citrus fruits composition. Adapted as an average from different studies [3-13]

Compound	Orange	Mandarin
Water	80.70	79.60
Sucrose	0.67	0.72
Glucose	1.48	1.73
Cellulose	4.23	4.40
Hemicellulose	1.84	1.88
Lignin	1.22	1.24
Pectin	3.70	3.92
Ash	0.77	0.79
Palmitic Acid	0.10	0.11
Stearic Acid	0.52	0.56
Oleic Acid	0.36	0.39
Linoleic Acid	0.05	0.07
Protein	1.76	1.88
D-limonene	0.37	0.40
Ascorbic Acid	0.30	0.32
Cryptoxhanthin	0.20	0.22
B-Carotene	0.64	0.68
Hesperidins	0.60	0.58
Naringin	0.25	0.27
Lactic Acid	0.04	0.04
Acetic Acid	0.05	0.05
Citric Acid	0.12	0.13
Propionic Acid	0.03	0.03
P-Coumaric Acid	0.00	0.00
Caffeic Acid	0.01	0.01
Ferulic Acid	0.00	0.00
Total	100.00	100.00

Note: Composition includes normalized values for components present in pulp, peel and seeds.

Table 5. Pulp, peel and seed percentages in citrus global composition

Fraction	Orange	Mandarin[a]
Pulp	70.44	68.98
Peel	27.58	28.57
Seed	1.97	2.45
Total	100.00	100.00

[a] Determined by experimental procedure.

On the other hand, from glucose can be obtained a great variety of products derived from the Kreps cycle. Therefore, the liquor rich in glucose is divided as follows: 20% of the glucose-rich liquor is used in the ethanol production plant. Although great volumes of alcohol for the oxygenation programs are required in Colombia, this is not the aim of this biorefinery. However, it satisfies the requirements of ethanol in other plants of the biorefinery such as pectin extraction, xylitol crystallization and lactic acid production. Still, an important fraction of ethanol can be recovered and commercialized. Another 20% of the glucose-rich liquor is used for the PHB production to supply the demand of biopolymers in Colombia. The remaining 60% of the glucose-rich liquor is separated in equal fractions (mass fraction) for the production of lactic acid and citric acid as high value-added products derived from sugars. Citric acid is produced to guarantee the requirements of this acid in the same biorefinery. Therefore, a fraction produced is integrated to the pectin extraction process. However, considering an internal mass integration, an interaction between streams is observed. In this way, an important fraction of citric acid is commercialized. Finally, the liquor remaining is sent toward the process of lactic acid production which is completely used to sale.

An important aspect to consider in the development of the biorefinery is the wastewater treatment to evaluate further scenarios resulting from the water recovery and recycle toward other plants. It is important to take into account the energy requirements for this scheme in which, also the cogeneration process of all solid waste obtained in different processes (lignin and cellular biomass from fermentations) is evaluated. Finally, the effect of mass and energy integration is considered in the scenarios description. Figure 1 shows the simplified process flowsheet for a citrus based biorefinery. Hereinafter, just for understanding the processing sections are named as plants.

2.2.1. Essential Oil Plant

Essential oil extraction from citrus peel is carried out using a supercritical fluid extraction (SFE) process. Extraction with supercritical fluids using carbon dioxide is very attractive because the solvent is not toxic and a green concept can be included. The process is carried out at low temperatures to avoid the thermal degradation of the compounds [14]. The dry and mill peel is feed into an extraction column in countercurrent flow with supercritical CO_2 (SC-CO_2). The process is carried out at 125 bar and 40ºC. The efficiency of the SFE process strongly depends on the pressure and temperature [14]. At low densities (<230 kg/m3) the selectivity between monoterpenes and oxygenated compounds is quite high but the oil solubility is low. Therefore, a high oxygenated concentration and recovery can be obtained, but high values of the solvent to feed ratio are required resulting disadvantageous from the economic point of view. The depressurization process is carried out in three separator column in order to minimize the essential oil lost due to rapid separation. The recovered CO_2 is recycled to the process and mixed with fresh CO_2.

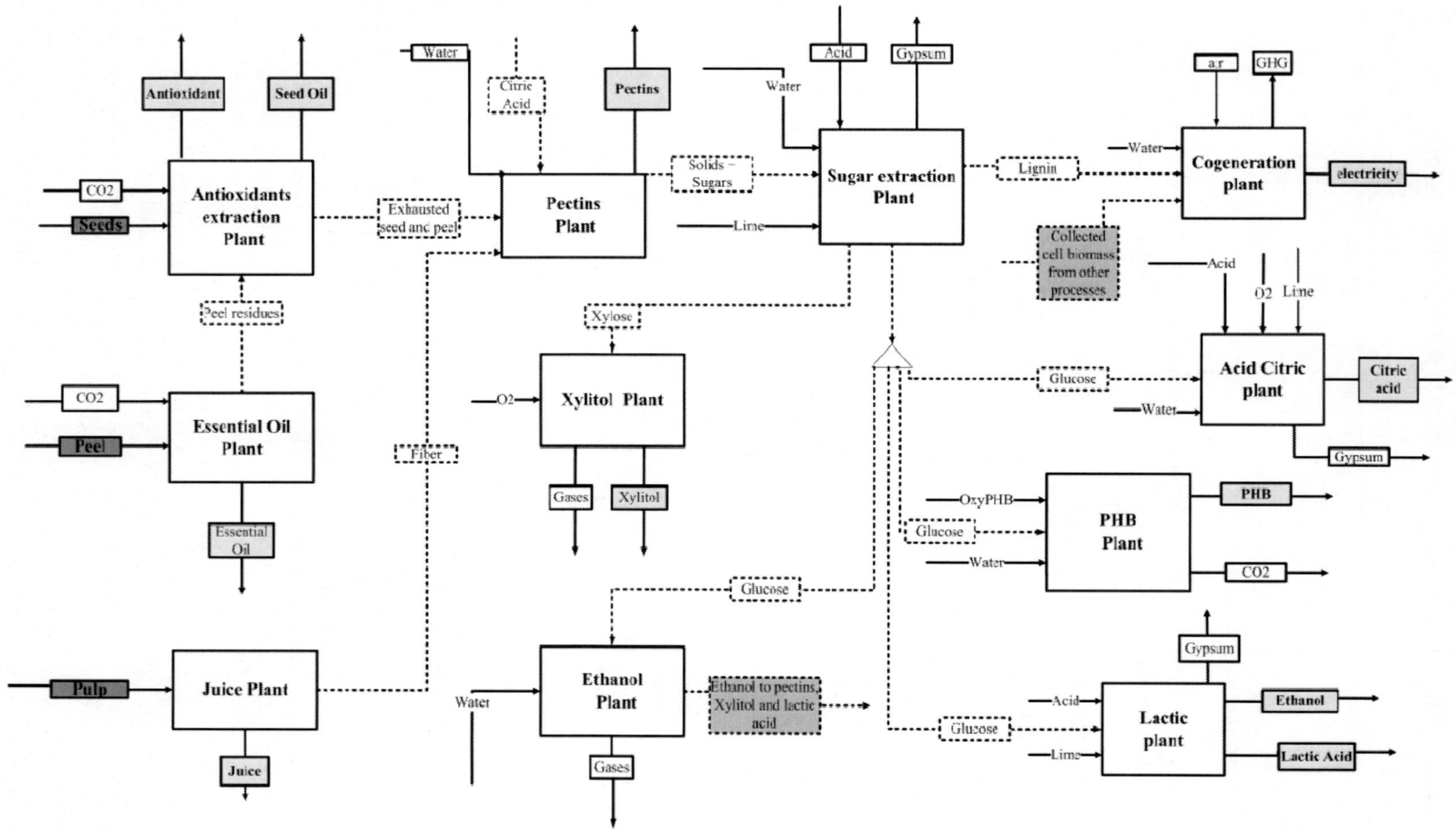

Figure 1. Simplified block process diagram for a citrus based biorefinery. Dot lines describe integrated streams. Yellow blocks represent products.

2.2.2. Antioxidant Plant

Antioxidants are obtained using SFE. Both the peel citrus waste from essential oil plant and citrus seed are used as feedstock. Firstly, the seed is dried and mixed with citrus peel waste. Then, the mixed stream is packed into an extractor unit with a 85% aqueous ethanol solution as co-solvent (CS). The SC-CO_2 is admitted into the system keeping the relation between the solvent mass (S) and the solid mass (F) constant and equal to 50. The process is carried out at 40°C (a reasonable value to preserve thermolabile compounds) and 350 bar. During the simulation, the process is carried out at 350 bar and 40°C to guarantee the supercritical conditions. Then, the CO_2 at the supercritical conditions is sent to the extractor at a solvent to solid ratio of 40:1 (CO_2/mass solid) [15]. The depressurization process is carried out at 20ºC and 1 bar in a separator column allowing the separation of CO_2 from the product-cosolvent mixture. The recovered CO_2 is recycled to the process and mixed with fresh CO_2. The extract obtained from the depressurization process is concentrated using ultrafiltration and is obtained as retentate rich in protein. The permeate stream is sent to nanofiltration to obtain an antioxidant rich fraction as retentate and oil seed as permeate [16].

2.2.3. Pectin Plant

The pectin process extraction is carried out by heating the dry and milled seed waste and fiber pulp of citrus in a reactor with acidified water (citric acid) at 80ºC. The pH of the process is adjusted at 2.5 [17]. The stream from the reactor is filtered to remove impurities and concentrated by evaporation 1/5 of the initial volume with a vacuum column. The pectin is precipitated from the solution by the addition of the equal reduced volume of 96% w/w ethanol. The coagulated pectin obtained is washed with ethanol 70% w/w to remove mono- and disaccharides. The resulting pectin is dried in a vacuum oven at 40ºC and milled. Finally, the ethanol resulted in evaporation step is recycled to the process and mixed with fresh ethanol [18].

2.2.3. Xylitol Plant

Xylitol is produced using the xylose-rich liquor from sugar plant which is preheated at 121ºC. After, the stream is inoculated with *Candida guilliermondii* and then oxygen is added to the fermentation process and the temperature is decreased. The xylitol is produced together with CO_2 which is separated in the top of the fermentor. Fermentation conditions are based on the work reported by Mussato, et al. [19]. After the fermentation, *C. guilliermondii* is filtered and the temperature is increased at 40ºC and a flash operation is used to concentrate the xylitol obtained. After evaporation, ethanol is added to decrease the solubility of xylitol and carry out crystallization at 5ºC [20]. Finally, the xylitol is centrifuged. The remaining ethanol is separated of the polyol molasses and it is recycled to the process after purge.

2.2.4. Citric Acid Plant

Citric acid is produced using 30% of the glucose-rich liquor obtained in the sugars plant in which a hydrolysis process is carried out. The glucose-rich hydrolysate is mixed with water and preheated at 121ºC and autoclaved and mixed with oxygen, then the temperature is decrease at 30ºC. Fermentation yields are based on the work from Ikram-ul, et al. [21]. The fermentation is carry out using *Aspergillus niger* which produced citric acid, CO_2 and hydrogen is small quantities, these last two components are separated from the fermentor while the remaining stream is preheated at 70ºC and the microorganism is filtered and separated. Proteins are precipitated from the stream containing citric acid at 70ºC and the remaining stream is mixed with calcium hydroxide to carried out the follow reaction an obtain calcium citrate at 95ºC:

$$3Ca(OH)_2 + 2C_6H_8O_7 \rightarrow Ca_3(C_6H_5O_7)_2 + 6H_2O$$

Calcium citrate is filtered and sent to a reactor to obtain citric acid again at 70ºC using sulfuric acid according to the following reaction:

$$3H_2SO_4 + Ca_3(C_6H_5O_7)_2 \rightarrow 2C_6H_8O_7 + 3CaSO_4$$

After, calcium sulphate is filtered. Water is evaporated from the citric acid stream and is recycled to the process while the citric acid is sent to a crystallization process at 38ºC, here calcium citrate is obtained in solid form and it is separate from citric acid by filtration. Finally, citric acid is dried. Downstream processing is based on description shown by Gluszcs and Ledakowicz [22].

2.2.5. Lactic Acid Plant

Lactic acid is obtained from 30% of the glucose-rich liquor obtained in sugars plant. Glucose is preheated at 121ºC and autoclaved. Fermentation process is carried out using *Lactobacillus delbrueckeii* at 37ºC. The products from this fermentation are preheated at 50ºC and the *L. delbrueckeii* is filtered and separated. Fermentation conditions were adapted from the work presented by Mussato, et al. [23]. Calcium hydroxide is used to obtain calcium lactate by following the reaction:

$$Ca(OH)_2 + 2C_3H_6O_3 \rightarrow CaC_6H_{10}O_6 + 2H_2O$$

Calcium lactate production is carried out at 50ºC and water is separated by filtration. Calcium lactate is mixed with ethanol from ethanol plant to decrease the solubility in water and to obtain lactic acid using sulfuric acid by following the reaction:

$$CaC_6H_{10}O_6 + H_2SO_4 \rightarrow CaSO_4 + 2C_3H_6O_3$$

After, the calcium hydroxide is filtered and separated from lactic acid. Then, majority of water and impurities are removed from lactic acid in ethanol solution in an ion exchange resin column. Then, the liquor is passed through an adsorption column packed with activated charcoal. In this step almost all water and impurities are removed. Lactic acid-rich liquor in ethanol is directed to a distillation tower obtaining lactic acid in the bottom at 0.998 in mass fraction and ethanol at 0.997 in mass fraction in the top. Downstream processing involves the combination of the procedures reported by Joglekar et al. [24] and Pal et al. [25].

2.2.6. Ethanol Plant

Fuel ethanol production process can be described in four stages according to reports made in previous work [26]: Pretreatment, hydrolysis, fermentation and separation and ethanol dehydration. Ethanol is obtained from 20% of the glucose-rich liquor from sugars plant. The stream of glucose and water is preheated at 121°C and autoclaved. Then fermentation is carried out at 35°C using *Saccharomyces cerevisiae*. After the fermentation stage, the culture broth containing approximately 7-10% (w/w) of ethanol is sent to the separation zone which consists of two distillation columns. From the fermentation, CO_2 is separated as well as the microorganism using a filter. In the first distillation column ethanol is concentrated nearly to 50-55% by weight. In the second column, the liquor is concentrated until the azeotropic point (96% wt) to be led to the dehydration zone with molecular sieves for obtaining an ethanol concentration of 99.7% by weight [27]. The main liquid effluent from the fuel ethanol process, the stillage, is evaporated up to a 30% concentration of solids. This procedure is done because there can be many problems in soil when all stillage volume is directly sprayed into the fields due to super saturation of phosphorus and other elements as reported previously for Colombia [28]. The remaining water from molecular sieve is mixed with the bottoms of the rectification column and with the water from stillage flashed. The bottoms from flashed stillage are mixed with *S. cerevisiae* separated after the fermentation.

2.2.7. PHB Plant

PHB is produced using 20% of the glucose-rich liquor from sugar plant. Glucose is diluted and preheated at 121°C and it is autoclaved. Then, oxygen is mixed and sent to a fermentation process at 37°C using *Cupriavidus necator* (*Ralstonia eutropha*) which is able to assimilate different carbon sources (in this case glucose) producing CO_2 and PHB, the fermentation conditions are based on previous studies [29]. Once the fermentation is done, the process follows a digestion which consists on cell lysis with chemical agents such as sodium hypochlorite assisted by temperature [30]. Once the biopolymer is extracted, residual biomass is separated by centrifugation and shipped as a solid residue for cogeneration systems. The resulting solution after centrifugation is washed in order to remove impurities to finally remove water by evaporation and spray drying to obtain almost pure PHB.

2.2.8. Cogeneration System

Cogeneration can be defined as a thermodynamically efficient way of use energy, able to cover complete or partially both heat and electricity requirements of a facility [31], [32]. Combined production of mechanical and thermal energy using a simple energy source, such as oil, coal, natural gas or biomass, has remarkable cost and energy savings, achieving operating also with a greater efficiency compared with systems which produce heat and electricity separately. If a cogeneration system is based on biomass or its residues, this system is known as biomass fired cogeneration. Most of biomass fired cogeneration plants are allocated at industrial sites guaranteeing a continuous supply of feedstock. Common examples are sugar and/or ethanol plants and paper mills. Additionally, the electricity surplus generated by biomass fired cogeneration projects could be used in rural areas close to production site, increasing the economic viability of the project [31].

For this study the technology used for cogeneration is the biomass integrated gasification combined cycle (BIGCC) [33]. Basic elements of BIGCC system include: biomass dryer, gasification chamber, gas turbine, heat steam recovery generator (HRSG) [34]. Gasification is a thermo-chemical conversion technology of carbonaceous materials (Coal, petroleum coke and biomass), to produce a mixture of gaseous products (CO, CO_2, H_2O, H_2, CH_4) known as syngas added to small amounts of char and ash. Gasification temperatures range between 875-

1275 K [35]. The Gas properties and composition of syngas changes according to the gasifying agent used (air, steam, steam-oxygen, oxygen-enriched air), gasification process and biomass properties. Syngas is useful for a broader range of applications, including: direct burning to produce heat and power or high quality fuels production or chemical products such methanol [36]. Heat recovery steam generator, is a high efficiency steam boiler that uses hot gases from a gas turbine o reciprocating engine to generate steam, in a thermodynamic Rankine Cycle. This system is able to generate steam at different pressure levels. According to process requirements a HSRG system can use single, double or even triple pressure levels. Each HSRG section is integrated by: Economizer, where cold water exits as saturated liquid. Evaporator where saturated steam is produced and superheated, where saturated steam is dried, overheating it beyond its saturation point. Citric bagasse and biomass from citric acid, ethanol, PHB, lactic acid and xylitol plants were used in cogeneration system.

2.3. Simulation Procedure

For a given scenario, flowsheet synthesis was carried out using process simulation tools. The objective of this procedure was to generate the mass and energy balances to calculate the raw materials, consumables, utilities and energy requirements. The main simulation tool used was the commercial package Aspen Plus v8.2 (Aspen Technology Inc., USA). Specialized package for performing mathematical calculations, especially for kinetic analysis, such as Matlab was also used. The fermentation stage for fuel ethanol production was calculated using the kinetic model reported by Rivera et al. [81]. The kinetic models used for hydrolysis steps were reported by Jin et al. and Rinaldi & Schüth [38, 82]. Kinetic model for detoxification were reported by Martinez et al. [83]. The kinetic model used for the calculation of PHB production was reported by Shahhosseini [84]. For some products (citric acid, xylitol, lactic acid and so on) fermentation conditions were adapted from different studies [86-88]. One of the most important issues to be considered during the simulation procedure was the appropriate selection of the thermodynamic models that describe the liquid and vapor phases. The Non-Random Two-Liquid (NRTL) thermodynamic model was applied to calculate the activity coefficients of the liquid phase and the Hayden-O'Conell equation of state was used for description of the vapor phase. For several biorefinery compounds an additional data for components was needed as the physical properties required for simulation were obtained from the work of Wooley and Putsche [89].

On the other hand, the methodology described by El-Halwagi [90] was used assessing the energy integration. Last allows a global view on the different levels of energy integration. Also the specialized package Aspen Energy Analyzer V8.2 served as the basis for the stream mapping of each scenario. This software was adapted to the Colombian context. On the other hand, mass integration strategies were based on the methodology of targeting, direct recycle and integration networks well discussed by El-Halwagi [90].

2.4. Cost Estimation

The estimation of the energy consumption was performed based on the results of the mass and energy balances generated by the simulation. Then, the thermal energy required in the heat exchangers, and re-boilers, as well as the electric energy needs of the pumps, compressors, mills and other equipments were calculated. The capital and operating costs were calculated using the software Aspen Economic Analyzer V8.2 (Aspen Technologies Inc.). Also energy consumption was determined regarding to different levels on energy integration and coupling cogeneration systems. On the other hand, specific parameters regarding to some Colombian conditions such as the raw material costs, income tax (33%), annual interest rate (16.02%) and labor salaries, among others, were incorporated in order to calculate the production costs per unit for the different obtained products. Table 6 shows prices for utilities and main raw materials and products in the different biorefinery configuration evaluated in the Colombian context. This analysis was estimated in US dollars for a 10-year period. The above-mentioned software estimates the capital costs of process units as well as the operating costs, among other valuable data, using the design information provided by Aspen Plus and data introduced by the user for specific conditions, as for example project location. The capital depreciation was calculated using the straight-line method. Equipment calculations were performed following the Aspen Economic Analyzer V.8.2 (Aspen Technologies, Inc., USA), which also uses correlations reported by Peters et al. [91]. Utilities, civil works, pipelines, man hours, and many different parameters were estimated using the same software.

2.5. Environmental Assessment

The Waste Reduction Algorithm WAR GUI, developed by the National Risk Management Research Laboratory de la U.S. Environmental Protection Agency (EPA) was used as the method for the calculation of the Potential Environmental impact. This method proposes to add a reaction of conservation over the potential environmental impact (PEI) based on impact of input and output flowrates from the process. The PEI for a given mass or energy quantity could be defined as the effect that those (energy and mass) will have on the environment if they are arbitrary discharged. The environmental impact is a quantity that cannot be directly measure; however, it can be calculated from different quantifiable measurable indicators. The WAR GUI software incorporates the Waste Reduction Algorithm in process design measuring eight categories. This categories are: Human toxicity by ingestion (HTPI), human toxicity by dermal exposition or inhalation (HTPE), aquatic toxicity potential (ATP), Global warming (GWP), Ozone depletion potential (ODP), Photochemical oxidation potential (PCOP) and acidification Potential (AP). This tool considers the impact by mass effluents and the impact by energy requirements of a chemical process, based on the energy and mass balances generated in Aspen Plus. Then the weighted sum of all impacts ends into the final impact per kg of products.

Other important environmental parameter is the Green House Gas (GHG) emissions associated to a chemical process. This procedure was completed following the IPPC Guidelines [103]. The GHG emissions are calculated using equivalent factors of 14 for CH_4, 4.5 for CO, 196 for NO_x. Also GHG emissions were calculated for energy needs when no integration with cogeneration schemes was done. The external energy source was charcoal with an emission factor of 94600 kg CO_2-e/TJ [103]. Nevertheless, when biorefinery configurations were coupled with cogenerations systems the GHG emissions corresponds to the mass balance in the exhausted gases from gasification/combustion.

Table 6. Prices for different products and raw materials used in the economic assessment

Item	Unit	Price	Reference
PHB	USD/kg	3.12	Posada et al. [93].
Anthocyanin	USD/kg	70 [a]	Sigma Aldrich [94].
Sugar	USD/kg	0.42	Asocaña [92].
Electricity	USD/kWh	0.1	National average.
High P. steam (105 bar)	USD/tonne	9.86	Moncada et al. [11].
Mid P. steam (30 bar)	USD/tonne	8.18	Moncada et al. [11].
Low P. steam (3 bar)	USD/tonne	1.57	Moncada et al. [11].
Water	USD/m3	1.252	National average.
Calcium Hydroxide	USD/kg	0.05	ICIS [95].
Sulfuric Acid	USD/kg	0.1	ICIS [95].
Orange Oil (Essential)	USD/kg	3.25	ICIS [95].
Citric Acid	USD/kg	1.74	ICIS[95].
Ethanol	USD/L	1.24	Fedebiocombustibles [96].
Mandarin Arrayana	USD/kg	0.78	Asohofrucol [97].
Orange Valencia	USD/kg	0.35	Asohofrucol [97].
Hexane	USD/L	0.31	ICIS [95].
CO_2	USD/kg	1.55	Cryogas Indura [98].
Pectin	USD/kg	12.62	ICIS [95].
Protein (feed)	USD/kg	0.21	ICIS [95].
Seed oil	USD/kg	0.96	ICIS [95].
Antioxidants	USD/kg	87.44 [c]	Sigma Aldrich [94]
Orange Juice	USD/kg	1.07	Ifu-Fruitjuice [99].
Xylitol	USD/kg	7.95	Xylitol Canada [100].
Lactic acid	USD/kg	1.87	ICIS [95].
Operator Labor	US$/h	2.14	Colombian Government
Supervisor Labor	US$/h	4.29	Colombian Government

[b]Sigma aldrich price for Pelargonidin chloride is 195.50 USD/mg. For this study due to a mixture of anthocyanins price was arbitrary selected to 70 US$/kg.

[c]Sigma aldrich average price for the antioxidant mixture is 3497.5 USD/kg. For this study due to a mixture of Phenolics price was calculated arbitrary using the 2.5% of total price (87.44 USD/kg).

Table 7. Biorefinery production volume per product.
Yield per ton of fruit

Productivity											
Sectors	E. Oil	Ant	S.Oil	Pect.	Elect.	C.Acid	PHB	L. Acid	Eth.	Xyl.	C. Juice
Production	T/day	T/day	T/day	T/day	MW	T/day	T/day	T/day	L/day	T/day	T/day
Sc 1 - Sc 4	7.91	8.65	21.13	75.04	0.00	13.34	10.56	37.56	14074.39	24.95	245.77
Sc 2 - Sc 5	7.91	8.65	21.13	75.04	0.00	13.34	10.56	37.56	14074.39	24.95	245.77
Sc 3- Sc 6	7.91	8.65	21.13	75.04	7.48	13.34	10.56	37.56	14074.39	24.95	245.77
Sc 7 - Sc 10	7.11	8.03	19.13	74.04	0.00	13.21	10.56	37.56	13974.39	24.68	252.18
Sc 8 - Sc 11	7.91	8.65	21.13	75.04	0.00	13.21	10.56	37.56	13974.39	24.68	252.18
Sc 9 - Sc 12	7.91	8.65	21.13	75.04	7.78	13.21	10.56	37.56	13974.39	24.68	252.18
Yield											
Sectors	E. Oil	Ant	S.Oil	Pect.	Elect.	C.Acid	PHB	L. Acid	Eth.	Xyl.	C. Juice
Unit/citrus T	kg/T	kg/T	kg/T	kg/T	MJ/T	kg/T	kg/T	kg/T	L/T	kg/T	kg/T
Sc 1 - Sc 4	3.30	3.60	8.80	31.27	0.00	5.56	4.40	15.65	5.86	10.40	102.40
Sc 2 - Sc 5	3.30	3.60	8.80	31.27	0.00	5.56	4.40	15.65	5.86	10.40	102.40
Sc 3- Sc 6	3.30	3.60	8.80	31.27	269.45	5.56	4.40	15.65	5.86	10.40	102.40
Sc 7 - Sc 10	2.96	3.34	7.97	30.85	0.00	5.51	4.40	15.65	5.82	10.28	105.08
Sc 8 - Sc 11	3.30	3.60	8.80	31.27	0.00	5.51	4.40	15.65	5.82	10.28	105.08
Sc 9 - Sc 12	3.30	3.60	8.80	31.27	280.08	5.51	4.40	15.65	5.82	10.28	105.08

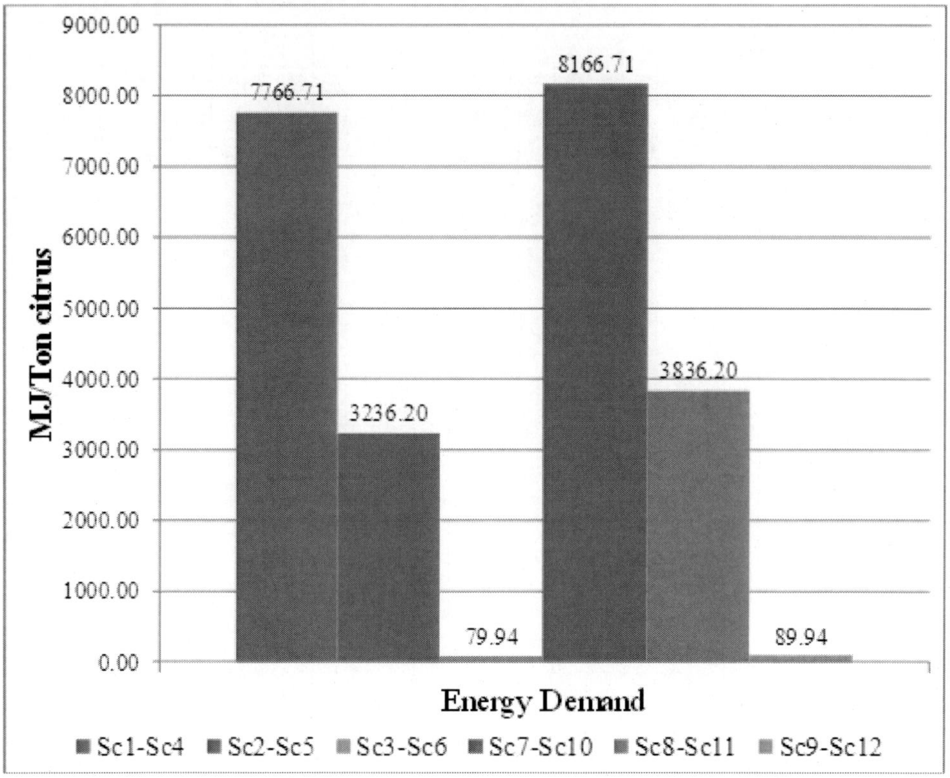

Figure 2. Energy requirements based on the different biorefinery heat integration levels for orange and mandarin.

3. Results

3.1. Technical Evaluation

Simulations were used to generate the respective mass and energy balances that were the basic input for the technical, economic and environmental analysis. Table 7 summarizes the production rates of the main product and it respective yield per ton of fruit (Orange and mandarin). This Table presents clearly that the production volumes do not change between the different scenarios to each fruit. The last fact is due to the fixed distribution above mentioned. However, for the scenarios that consider energy cogeneration, electricity appears as a new product. From the production volume point of view, the difference between orange (sc 1 to sc 6) and mandarin (sc 7 to sc 12) is important because of the variations in the global composition.

On the other hand, the biorefineries assessment is centered in the influence of the energy and mass integration. Therefore, one of the priorities in the evaluation is the determination of the energy requirements according to the scenarios description. Considering this, Figure 2 shows the energy requirements of the three levels of integration for both raw materials. This Figure presents the gradual decrease of total energy consumption. This situation represents a decrease of costs and environmental impact due to the external fuel used. Data showed in Figure 2 corresponds to the required energy for heating and cooling fluids, however, doing the corresponding energetic integration, the targets are determine over the reductions on heating and cooling utilities. For the orange case, when a scenario with full energy integration without cogeneration appears, the total saving in energy needs is 96.72 and 41.76% for cooling and heating utilities, respectively. For the mandarin case, the saving consists in 94.72 and 43.73% for cooling and heating utilities, respectively, when full energy integration is present without cogeneration. Due to these greater savings an important decrease is observe in the total energy requirements.

The saving percentages were determined finding the optimum pinch point that allows the integrations between different process streams and take advantage of the generated energy in the process. Figure 3 presents the composite curves for the heating and cooling utilities for the configuration based on orange. The mandarin scheme is not present because the integration levels are very similar to the orange case. Finally, this result is due to the similar processing conditions for both fruits, because its distribution and technologies are the same.

The influence of the energy cogeneration savings is reflected directly over the heating utilities. In addition to the electricity generated, also high, mid and low pressure steam was produced according to the plant requirements. Once the systems are energetically integrated, the remaining energy required is covered by the steam generated in the cogeneration system. This helps to demonstrate the important influence on process integrations in term of energy in sustainable biorefineries design. Therefore, the decrease of energy requirements is justified as is showed in Figure 2 with full energy integration coupled with the cogeneration system.

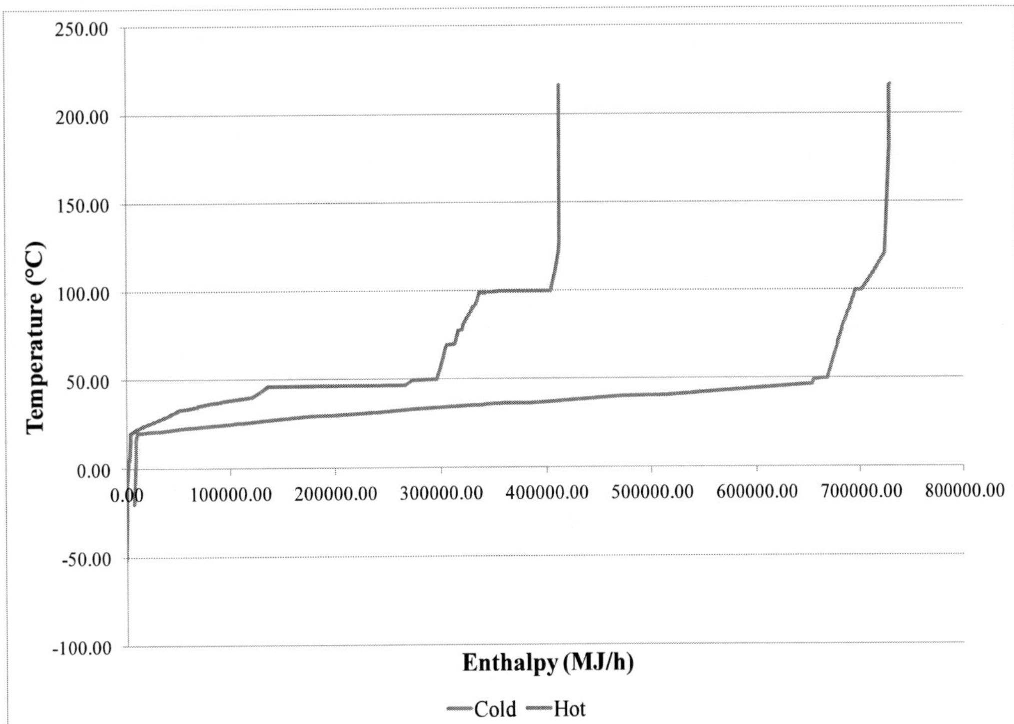

Figure 3. Composite curves for heating and cooling streams of the configuration based in orange.

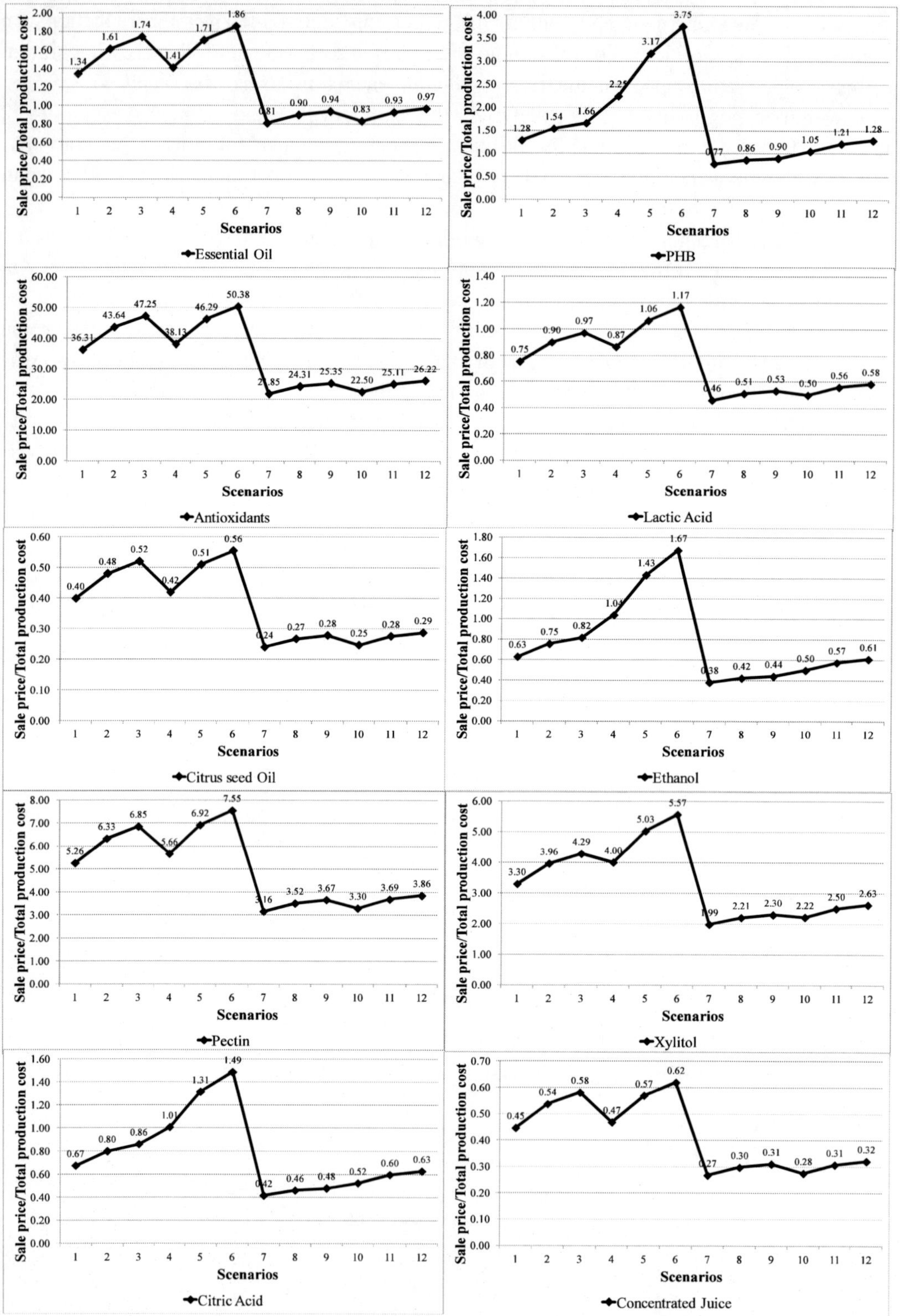

Figure 4. Sale Price to total production cost ratio per product for all scenarios.

These scenarios have impact over the environmental and economic evaluations which will be explained later. On the other hand, the mass integration is reflected over the decrease of both, fresh sources consumption and environmental impact due to the treatment of the organic charge in the leaving waters of the process. The focus of the mass integration permits the inclusion of the recovery of the water present in the fruits.

As observed in Table 4, the percentage of water in these fruits is very high (approximately 80%). Therefore, one of the discussions over the biotechnological process is the great volumes of water resources used to the conditioning of the substrates to reach the best conditions in the fermentations and/or reaction processes. This is the reason why the stillage are concentrated and the residual water from the separation processes of each biotechnological product is recovered at a high possible level.

3.2. Economic Evaluation

To determine the influence of the mass and energy integration on the economic evaluation, the sale price to total production costs ratio is shown for each product in the biorefineries (Figure 4). The impact on the production cost because of energy integration occurs in the utility cost reduction, whereas the influence of mass integration applies in a drop of the raw material costs. The influence of capital cost on the total production cost is not significant because the increase in the mass and energy integration is distributed over all plants in the process. In Figure 4, for each product, a significant increase in the sale price to production cost ratio is observed as the level of integration increases. For all products the scenario that most increases this ratio is the one with mass and energy integration plus cogeneration.

Elsewhere, on the Figure 4 is clear that the relationship of sales on product cost is lower for mandarin-based scenarios. This is due to the high prices of mandarin in Colombia due to the low level of domestic production. Similarly Figure 4 shows that there are products that have a sale to cost ratio lesser than the unit (Concentrated juice, citrus seed oil, ethanol for some scenarios, lactic acid for some scenarios, and mandarin essential oil for all the scenarios, among many others), meaning that as single products are not profitable. But other high value-added products such as the case of antioxidants, xylitol, pectin and others help to subsidize these processes and products. For all scenarios the most valuable product are the antioxidants.

Considering the above mentioned, the direct relationship between each process is essential in the design of biomass based refineries. So to show the overall economic performance for all the scenarios, Figure 6 shows the profit margin for each one. Last is calculated as the ratio between the sales of all products and the total production cost in the biorefinery. As expected, the behavior is very similar to that for individual products shown in Figure 4. The higher profit margins are obtained in scenarios with the highest integration levels. However, a notable decrease in the profit margin when using mandarin as feedstock compared to orange is observed. Last is due to the high cost of mandarin as explained above. In order to show the contribution on sales of each product, Figure 7 shows the percentages for scenarios based on orange and mandarin with and without cogeneration.

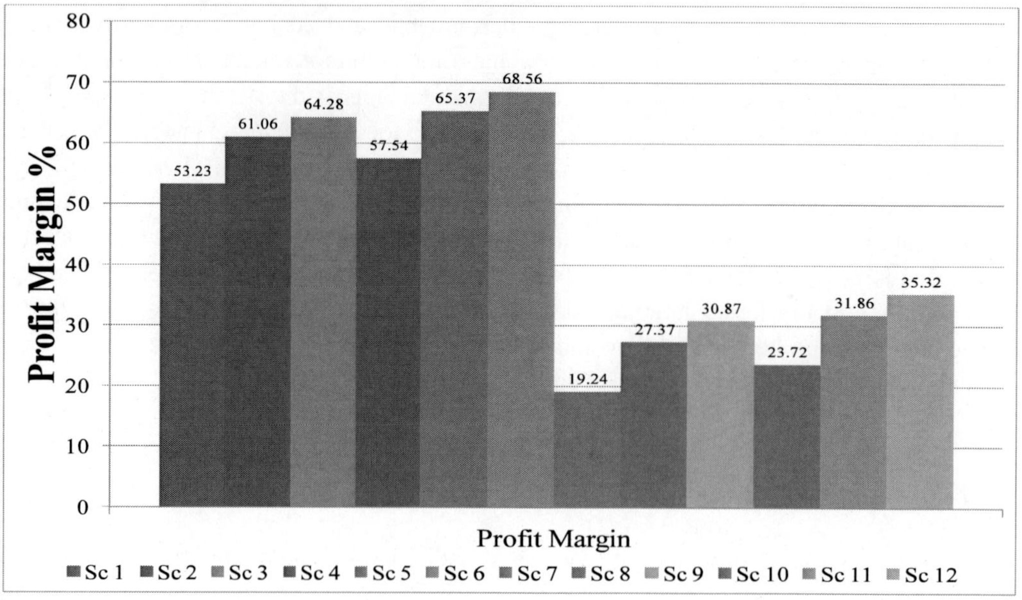

Figure 5. Profit Margin per scenario for the different biorefinery configurations.

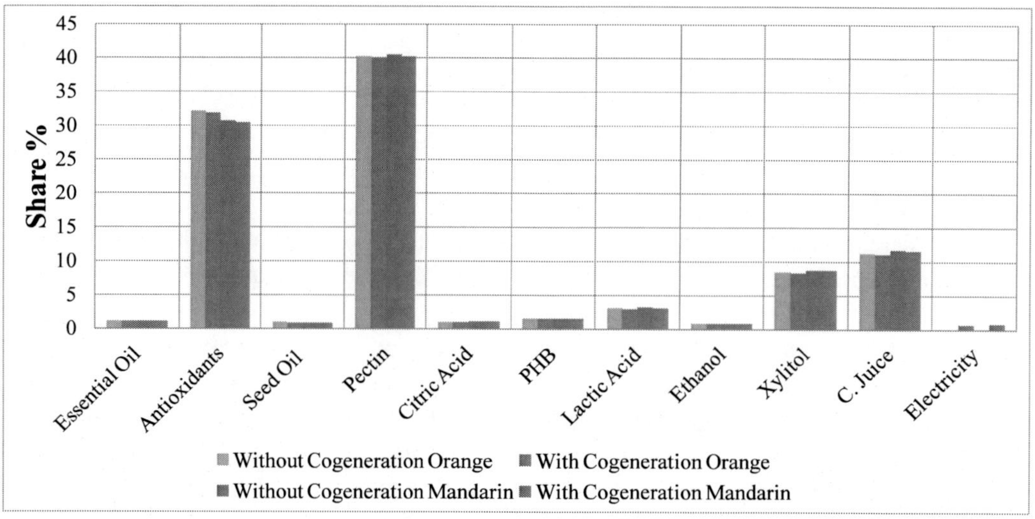

Figure 6. Percentage on sales per product for the different citrus based biorefineries configuration.

3.3. Environmental Evaluation

Another important aspect to consider in the design of sustainable processes is the potential environmental impact. As mentioned before, the energy influence is reflected in the generation of greenhouse gases emissions, which holds interests in the potential environmental impact. At this point it is very important to consider that the energy source to generate steam when no heating or coupling of energy integration with the cogeneration system is coal. The emission factor presented in kg CO_2-e per TJ of coal is 94600 as the reported in the IPCC guidelines [38].

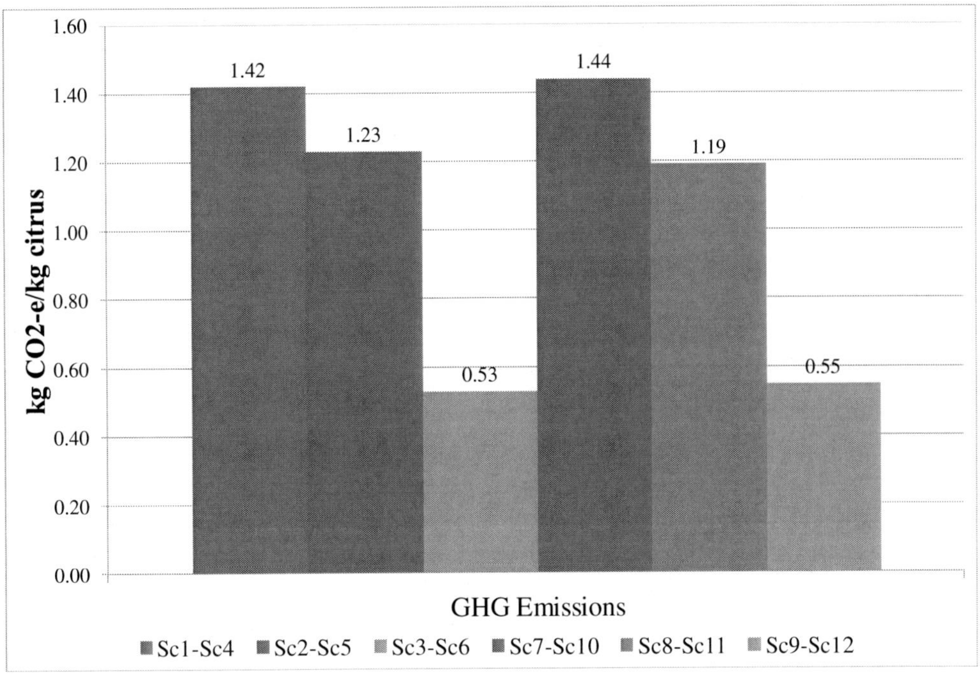

Figure 7. GHG emissions per scenario represented as kgCO2-e/kg citrus.

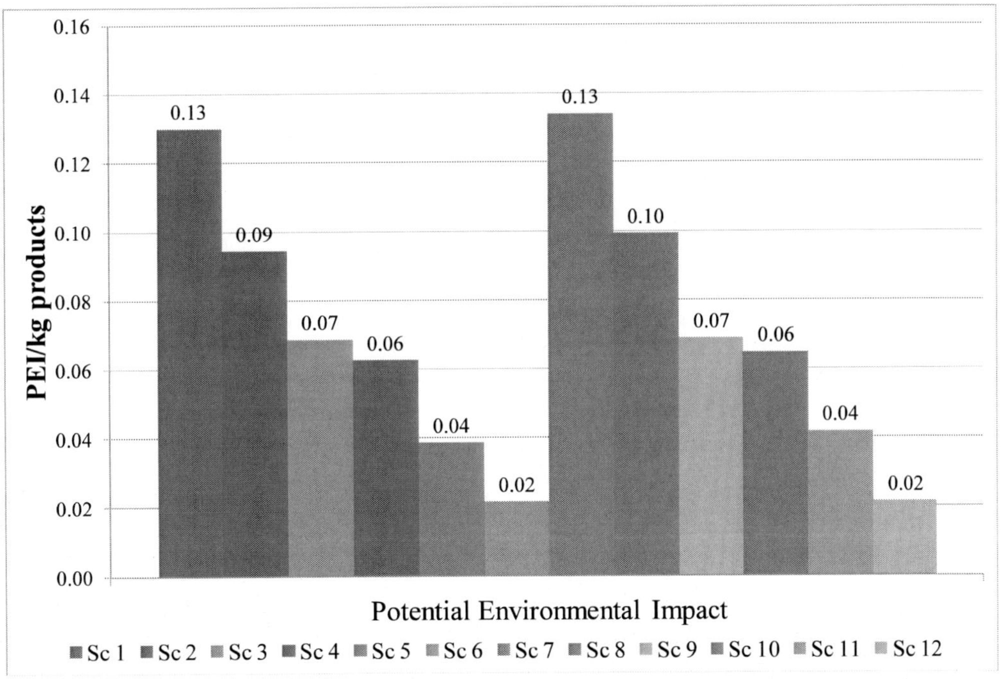

Figure 8. PEI per kilogram of products for all scenarios in a citrus based biorefinery.

On the other hand, in many cases, the potential environmental impact generated by the effluent water is due to the high organic content effluents may have. Considering this, for each scenario the GHG emissions and the total potential environmental impact per kilogram of products is presented as the environmental evaluation results. In this way, Figure 7 shows kg CO_2-e per kg of fruit. This analysis also considered emissions from the material balance as the case of fermentations that release carbon dioxide. At this point it is important to note that for scenarios coupled with cogeneration systems, the GHG emissions corresponds to the exhausted gas composition leaving the process.

As it was expected, the scenarios that do not count with energy integration are those which require higher energy consumption. Also for the outside selected source for energy (coal) the CO_2 equivalents was increased in comparison to the use of biomass. Last can be reflected in the integration patterns with high energy level coupling with cogeneration schemes.

Furthermore, the potential environmental impact is affected by both energy levels and process waste streams final disposition. This is why it is observed in Figure 8 that as the level of global integration increases the potential environmental impact decreases.

In this chapter, it is evident that the use of strategies such as mass and energy integration in hierarchical levels can increase significantly the economic viability when a multiproduct portfolio is presented. Therefore, the inclusion of these strategies in the design of a sustainable biorefinerie permits to show a significant reduction in the potential environmental impact. Moreover, the inconvenient presented in processes for obtaining new products is due to the energy consumption, the low integration and low used of the energy from process streams and important logistic problems when separated technologies and processes are no integrated into one biorefinery.

CONCLUSION

The best scheme of the process corresponds to the scenarios 6 and 12 for orange and mandarin, respectively. On the other hand, it is important to highlight the possible social impact that can have the generation of these projects which involved an extension and development of the rural sectors. In the same way, the food security preservation is highlighted as social preoccupation inside the design of the sustainable biorefineries such is the case of the juice, antioxidant, oils and pectin production, among others.

REFERENCES

[1] Ladaniya, M.S., *Introduction. Citrus fruit production and prospects., in Citrus fruit: biology, technology and evaluation,* M.S. Ladaniya, Editor 2008, Academic Press. Elsevier: UK.

[2] MinAgricultura, *Anuario Estadistico del sector agropecuario y pesquero.* Ministerio de agricultura y desarrollo rural, 2010.

[3] Ortiz, J.M., Botany: taxonomy, morphology and physiology of fruits, leaves and flowers. Citrus: the genus citrus. *Medicinal and aromatic plants--industrial profiles*, 2002. 26: p. 16-35.

[4] Marín, F.R., et al. By-products from different citrus processes as a source of customized functional fibres. *Food Chemistry*, 2007. 100(2): p. 736-741.

[5] Ladaniya, M.S., Citrus fruit: biology, technology and evaluation, ed. C.F. Biochemistry2008: Academic Press.

[6] Ángel Siles López, J., Q. Li, and I.P. Thompson, Biorefinery of waste orange peel. *Critical reviews in biotechnology*, 2010. 30(1): p. 63-69.

[7] Grohmann, K., R. Cameron, and B. Buslig, Fractionation and pretreatment of orange peel by dilute acid hydrolysis. *Bioresource technology*, 1995. 54(2): p. 129-141.

[8] Ververis, C., et al. Cellulose, hemicelluloses, lignin and ash content of some organic materials and their suitability for use as paper pulp supplements. *Bioresource technology*, 2007. 98(2): p. 296-301.

[9] Okoye, C., C. Ibeto, and J. Ihedioha, Preliminary Studies on the Characterization of Orange Seed and Pawpaw Seed Oils. *American Journal of Food Technology*, 2011. 6: p. 422-426.

[10] Saloua, F., N.I. Eddine, and Z. Hedi, Chemical composition and profile characteristics of Osage orange Maclura pomifera (Rafin.) Schneider seed and seed oil. *Industrial Crops and Products*, 2009. 29(1): p. 1-8.

[11] Akpata, M. and P. Akubor, Chemical composition and selected functional properties of sweet orange (Citrus sinensis) seed flour. *Plant Foods for Human Nutrition (Formerly Qualitas Plantarum)*, 1999. 54(4): p. 353-362.

[12] Anwar, F., et al. Physico-chemical characteristics of citrus seeds and seed oils from Pakistan. *Journal of the American Oil Chemists' Society,* 2008. 85(4): p. 321-330.

[13] Boluda-Aguilar, M., et al. Mandarin peel wastes pretreatment with steam explosion for bioethanol production. *Bioresource technology,* 2010. 101(10): p. 3506-3513.

[14] Gironi, F. and M. Maschietti, Continuous countercurrent deterpenation of lemon essential oil by means of supercritical carbon dioxide: Experimental data and process modelling. *Chemical Engineering Science,* 2008. 63: p. 651-661.

[15] Cerón, I.X., J.C. Higuita, and C.A. Cardona, Design and analysis of antioxidant compounds from Andes Berry fruits (Rubus glaucus Benth) using an enhanced-fluidity liquid extraction process with CO2 and ethanol. *The Journal of Supercritical Fluids*, 2012. 62(0): p. 96-101.

[16] Carvalho Jr, R.N., et al. Supercritical fluid extraction from rosemary (Rosmarinus officinalis): Kinetic data, extract's global yield, composition, and antioxidant activity. *The Journal of Supercritical Fluids*, 2005. 35(3): p. 197-204.

[17] Liu, Y., J. Shi, and T.A.G. Langrish, Water-based extraction of pectin from flavedo and albedo of orange peels. *Chemical Engineering Journal,* 2006. 120(3): p. 203-209.

[18] Yeoh, S., J. Shi, and T.A.G. Langrish, Comparisons between different techniques for water-based extraction of pectin from orange peels. *Desalination*, 2008. 218(1–3): p. 229-237.

[19] Mussatto, S.I. and I.C. Roberto, Establishment of the optimum initial xylose concentration and nutritional supplementation of brewer's spent grain hydrolysate for xylitol production by Candida guilliermondii. *Process Biochemistry*, 2008. 43(5): p. 540-546.

[20] Vyglazov, V., Kinetic characteristics of xylitol crystallization from aqueous-ethanolic solutions. *Russian journal of applied chemistry*, 2004. 77(1): p. 26-29.

[21] Quintero, J.A., et al. Fuel ethanol production from sugarcane and corn: comparative analysis for a Colombian case. *Energy*, 2008. 33(3): p. 385-399.

[22] Gluszcs, P. and S. Ledakowicz, Chapter 9. Downstream Processing in Citric Acid Production, in Citric Acid Biotechnology, B. Kristiansen, M. Mattey, and J. Linden, Editors. 2002, Taylor & Francis: USA, UK.

[23] Mussatto, S.I., et al. Effects of medium supplementation and pH control on lactic acid production from brewer's spent grain. *Biochemical Engineering Journal*, 2008. 40(3): p. 437-444.

[24] Joglekar, H., et al. Comparative assessment of downstream processing options for lactic acid. *Separation and purification technology*, 2006. 52(1): p. 1-17.

[25] Pal, P., et al. Process intensification in lactic acid production: A review of membrane based processes. *Chemical Engineering and Processing: Process Intensification*, 2009. 48(11): p. 1549-1559.

[26] Cardona, C.A. and Ó.J. Sánchez, Fuel ethanol production: Process design trends and integration opportunities. *Bioresource technology*, 2007. 98(12): p. 2415-2457.

[27] Pitt Jr, W., G. Haag, and D. Lee, Recovery of ethanol from fermentation broths using selective sorption–desorption. *Biotechnol. Bioeng.*, 1983. 25(1): p. 123 - 131.

[28] Quintero, J.A., et al. Fuel ethanol production from sugarcane and corn: Comparative analysis for a Colombian case. *Energy*, 2008. 33(3): p. 385-399.

[29] Shahhosseini, S., Simulation and optimisation of PHB production in fed-batch culture of Ralstonia eutropha. *Process Biochemistry*, 2004. 39(8): p. 963-969.

[30] Herron, J.S., J.D. King, and D.C. White, Recovery of poly-β-hydroxybutyrate from estuarine microflora. *Applied and environmental microbiology*, 1978. 35(2): p. 251-257.

[31] Nagel, J., Determination of an economic energy supply structure based on biomass using a mixed-integer linear optimization model. *Ecological Engineering,* 2000. 16, Supplement 1(0): p. 91-102.

[32] Prasad, S.B., Biomass-fired steam power cogeneration system: A theoretical study. *Energy Conversion and Management*, 1995. 36(1): p. 65-77.

[33] Balat, M., et al. Main routes for the thermoconversion of biomass into fuels and chemicals. Part 2: Gasification systems. *Energy Conversion and Management*, 2009. 50(12): p. 3158 - 3168.

[34] Larson, E.D., R.H. Williams, and M.R.L.V. Leal, A review of biomass integrated-gasifier/gas turbine combined cycle technology and its application in sugarcane industries, with an analysis for Cuba. *Energy Sustain. Develop.,* 2001. 5(1): p. 54 - 76.

[35] Ahmed, I.I. and A.K. Gupta, Sugarcane bagasse gasification: Global reaction mechanism of syngas evolution. *Applied Energy*, 2012. 91(1): p. 75-81.

[36] Xu, Y., et al. High efficient conversion of CO2-rich bio-syngas to CO-rich bio-syngas using biomass char: a useful approach for production of bio-methanol from bio-oil. *Bioresource technology*, 2011. 102(10): p. 6239-6245.

[37] Moncada, J., M.M. El-Halwagi, and C.A. Cardona, Techno-economic analysis for a sugarcane biorefinery: Colombian case. Bioresource technology, 2012. In Press.

[38] Eggleston, H., 2006 IPCC Guidelines for National Greenhouse Gas Inventories. *Forestry,* 2006. 5(OVERVIEW): p. 1-12.

In: Citrus Fruits
Editor: Daphne Simmons

ISBN: 978-1-63484-078-1

Chapter 2

EXPLORING BIOACTIVITY OF HESPERIDIN, NATURALLY OCCURRING FLAVANONE GLYCOSIDE, ISOLATED FROM ORANGES

***Ljubica Tasic*[1*], *Boris Mandic*[1,2], *Caio H. N. Barros*[1], *Daniela Z. Cypriano*[1], *Danijela Stanisic*[1], *Lilian G. Schultz*[1], *Lucimara L. da Silva*[1], *Mayra A. M. Mariño*[1] *and Verônica L. Queiroz*[1]**

[1]Chemical Biology Laboratory, Institute of Chemistry, Organic Chemistry Department, State University of Campinas, Campinas, SP, Brazil

[2]Faculty of Chemistry, University of Belgrade, Belgrade, Serbia

ABSTRACT

Hesperidin is the principal bioflavonoid found in citrus fruits, with very interesting bioactivity properties that still are the object of intensive research. Hesperidin and its aglycon form, hesperetin, are present in large quantities in oranges (*Citrus sinensis*) in particular. In young, immature oranges, these flavones can account for up to 14% of the fruit weight. Hesperetin is the 3',5,7-trihydroxy-4'-methoxy flavanone, and hesperidin contains not only the flavanone moiety, but also a rutinose disaccharide that has one D-rhamnose united with glycoside bond to the D-glucose unit. This paper presents and discusses chemical and physical properties of the orange bioflavones, as well as the most common methods of isolating and purifying these compounds. As a secondary plant metabolite, hesperidin is produced as a protective agent in citrus, and its defense role and biosynthesis will also be briefly discussed here. Many interesting bioactive properties of these phytochemicals have been reported, including antioxidant, anti-inflammatory, hypolipidemic, vasoprotective and anticarcinogenic properties, and an extensive review of these properties will be presented. Last but not least, we will present the most up-to-date developments in the research field that account for the mechanisms of action of these compounds.

* Corresponding author: Ljubica Tasic. E-mail: ljubica@iqm.unicamp.br.

1. INTRODUCTION

Due to the numerous and very interesting properties of their constituents, aromatic and medicinal plants have been gaining a great deal of attention in academia and health sciences, as well as in industries such as the agrarian, chemical, and food industries. The flavor, aroma, and pharmaceutical properties of these plants are being explored in certain cosmetics, fragrances and drugs [1]. In this context, citrus fruits are especially important, not only because they are used as food, mainly in the form of juice, but also because of their essential oil and other active compounds which have important medicinal and cosmetic properties. Orange, tangerine, lemon, and grapefruit are fruits that belong to the family of *Rutaceae*, gender *Citrus*. The sweet orange (*Citrus sinensis*) is the most popular citrus fruit, widely consumed around the world. The prevention of scurvy (a disease caused by lack of vitamin C in our organism [2, 3]) was possibly the first medical use of orange, predating contemporary scientific understanding of the benefits that consuming oranges brings to the human organism.

Many different compounds can be isolated from *Citrus sinensis*. Ethanol extract of *Citrus sinensis* peel is composed mainly from carbohydrates, alkaloids, saponins, tannins, oil, steroids, and phenols. Orange juice and peel also contain numerous flavonoids [1]. Flavonoids are the secondary plant metabolites and represent a large proportion of the polyphenols found in seed, cortex, pulp, leaf, and flowers of various plant species, fruit, and herbs [4]. They can be regarded as C6-C3-C6 compounds, in which each C6 moiety is a benzene ring. Variations in the state of oxidation of the connecting C3 moiety determine the properties and class of such compounds. These compounds can be divided into six subgroups, as illustrated in Figure 1: (a) flavanone, (b) flavan-3-ol, (c) flavone, (d) flavone-3-ol, (e) anthocyanins, and (e) isoflavones.

These specific structures give these compounds a wide and effective range of uses in many areas [2]. Several bioactive properties of these phytochemicals have been reported, including antioxidant, anti-inflammatory, anticancer, chemo-protective, antimicrobial, and antiviral properties [5].

Most of the flavanones from Citrus species are found glycosidazed and belong to two groups: rutinosides and neohesperidosides [1]. Hesperidin is a flavanone rutinoside, i.e., its aglycone hesperetin is glycosidazed by rutinose. As a one of the bioflavonoids found in large quantities in the exocarp and endocarp of Citrus species, hesperidin is the major active constituent of tangerine (*Citrus reticulata*) and sweet orange (*Citrus sinensis*) peels. In young, immature oranges it can account for up to 14% of the fresh weight of the fruit. It is proposed, following *in vitro* studies, that hesperidin plays a role in plant defense and acts as an antioxidant.

The orange juice industry by-products mentioned above are only sporadically produced, and sometimes neglected or forgotten.

This chapter is focused on these by-products, and in particular on hesperidin, which has very interesting bioactive properties that are still the object of intensive research. Isolation and purification of hesperidin and hesperetin, and the bioactivities and mechanisms of action of these phytochemicals, are presented. In addition, a number of other high value-added products that can be obtained or extracted from orange peel, such as nanocellulose and bioethanol, are also discussed.

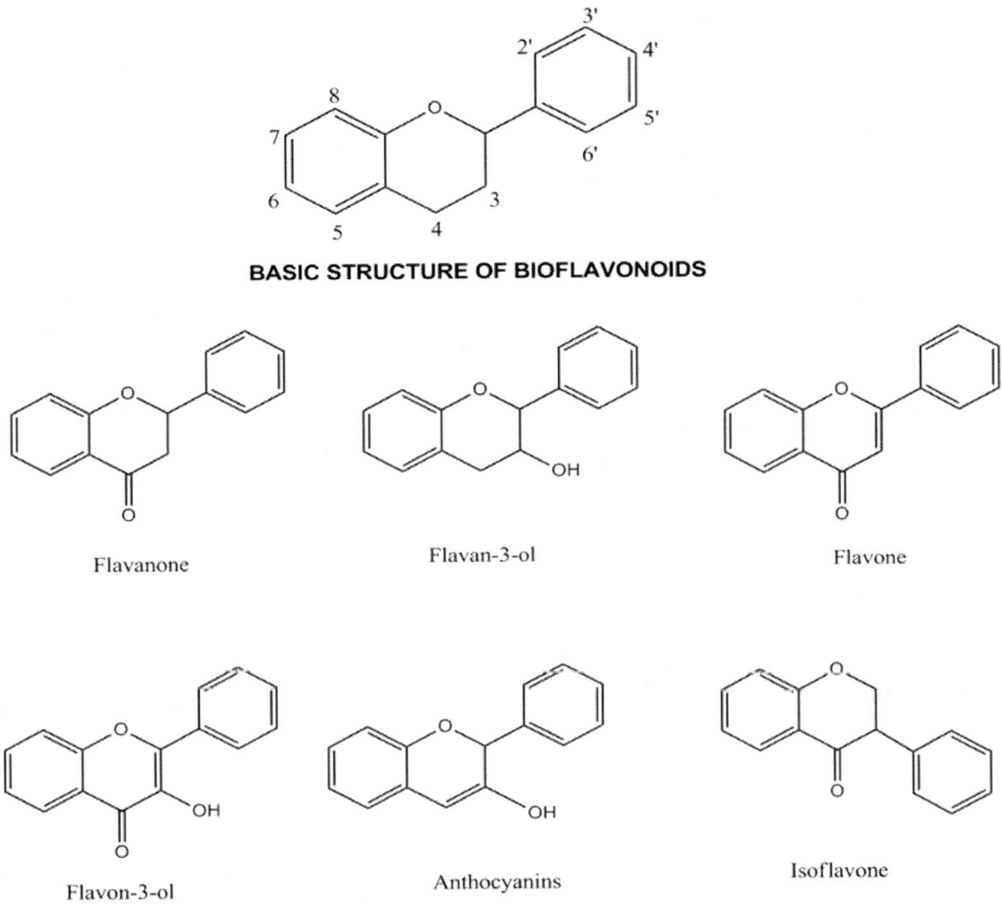

Figure 1. Basic structure of bioflavonoids and skeletons of their major subgroups (chemical formula: C6-C3-C6): (a) flavanones, (b) flavanes-3-ol, (c) flavones, (d) flavones-3-ol, (e) anthocyanins and (e) isoflavones.

2. Citrus Plants and Fruit

The *Citrus sinensis* tree is not very tall, and has a conical top and many branches often with thick spines. Orange leaves are dark green, glossy, and smooth, and smell similar to the fruit. Flowers are small, waxy, and greenish-white, with a specific fragrance and broad, saucer-shaped calyx. The species of the specific Chinese origin, *C. sinensis*, starts flowering and bearing fruit after 3-5 years. Trees are hermaphroditic and insect-pollinated [3]. The fruit consists of a leathery, tightly adherent peel that protects the juicy inner pulp, which is divided into segments that may or may not contain seeds, depending on the cultivar.

Fruit of the Citrus species is composed of pericarp and can be divided into the exocarp (the external part of the peel (flavedo)), mesocarp, (the white internal part of the peel (albedo)) and endocarp (Figure 2).

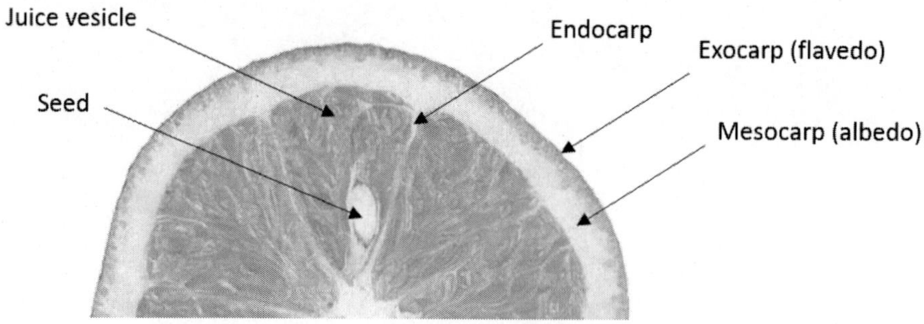

Figure 2. Illustration of an orange fruit slice cut into two halves, showing the morphology of the orange fruit with seed in the centre of the juice vesicle and endo-, meso- and exocarps.

The layers of flavedo tissue (cuticle) are covered with epicuticular wax layers formed from alcohols, higher *n*-alkanes, aldehydes, ketones, and other compounds, which are responsible for restricting the water loss from the peel. Albedo is the white part of the peel, which consists of cellulose, hemicellulose, and pectin.

Endocarp is divided into 8-16 carpels clustered around and joined to the floral axis, forming the core of the fruit. The carpels form segments, called locules, the sacs in which the seeds and juice grow. Hesperidin is present in all segments of the fruit [5, 6].

Growth and maturity of the Citrus fruit is dependent on various factors, such as climate conditions most importantly, as well as soil type, water availability, cultural practices, and nutrient supply. Climatic factors such as temperature, relative humidity, rainfall, wind velocity, and days with high sun indexes have significant effects on fruit morphology and ingredients. Therefore, cultivation of Citrus fruit over centuries has shown that subtropical climates (Sao Paulo, Florida, California, and Mediterranean) are better for the quantity, quality and nutrient content of the fruit than other climate zones. The best quality fruit is grown in Mediterranean-type climates, characterized by relative dryness (low rainfall), hot summers, and cool, wet winters [5].

Two stages of slow growth with a period of rapid growth in between are three main stages that define the growth of Citrus fruit. In the first stage, which lasts for 30 to 40 days, the cuticle is not developed despite intense cell division, and the small fruitlets are vulnerable and sensitive to superficial damage by wind and insects. Brazil, the US (Florida), and India experience the biggest problems at this stage due to the rains and abundance of insects in the post-bloom period. The second stage is determined by juice forming and developing of their distinctive solutes, which are high in organic acid content and low in sugar content. As the orange matures, the sugar percentage increases while acids decline, and the fruit also has a high water content (80-90%). Changes during the last, slow stage influence maturity. Carotenoid, epicuticular wax, ethylene and sugar content increase, and the chlorophyll percentage reduces to a minimum. Also, fruit weight decrease due to the respiratory pattern and juice acids decrease during this stage. Ascorbic acid content increases during the first stage and decreases with the progress of maturation, while the flavonoids content is increased during the first stage and remains constant [5].

3. Hesperidin and Hesperetin: Structure, and Chemical and Physical Properties

Hesperidin is a bioflavonoid that belongs to the flavanones class. It is glycoside (hesperetin 7-rutinoside) in which aglycone hesperetin (3',5,7-trihydroxy-4'-methoxy flavanone) is glycosidased by rutinose, disaccharide that has one D-rhamnose united by a glycoside bond to the D-glucose unit (Figure 3) [7]. Its molecular formula is $C_{28}H_{34}O_{15}$ and molecular weight 610.57 a.u.

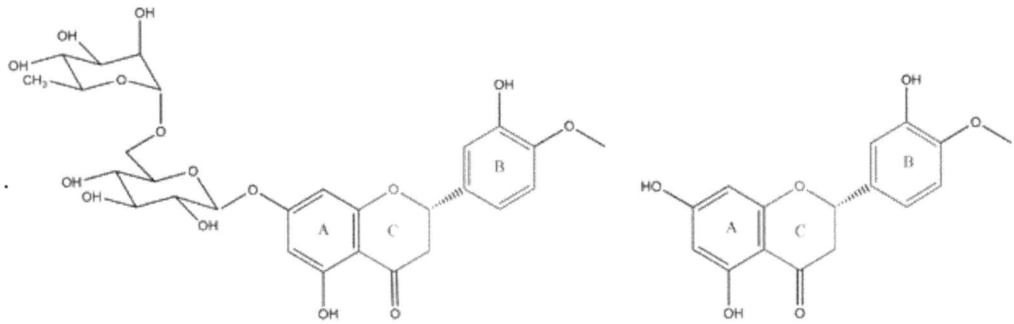

Figure 3. Chemical structure of hesperidin (hesperetin-7-O-β-rhamnoglucoside) and its aglycone form (3',5,7-trihydroxy-4'-methoxy flavanone).

Electron micrographs of hesperidin crystals at different scales are presented in Figure 4. The melting point of hesperidin ranges from 252° to 254°C [8]. It is easily soluble in dilute alkali and in pyridine giving a clear yellow solution. Hesperidin is also slightly soluble in methanol and hot glacial acetic acid, and almost insoluble in acetone, benzene, and chloroform [9].

Indication of the hesperidin presence in a sample of orange peel can be performed by means of a ferric chloride test. This test consists of addition of ferric chloride solution to a sample, and if hesperidin is present, a wine red color is produced. Hesperidin presence can also be detected with a magnesium-hydrochloric acid reduction test using dropwise addition of concentrated hydrochloric acid to an ethanol solution of hesperidin containing magnesium, resulting in a bright violet color occurrence [4].

a b c

Figure 4. Scanning electron micrographs of hesperidin crystals (needles) at different scales, (a) 10 μm (b) 5 μm and (c) 1 μm.

4. HESPERIDIN EXTRACTION

As mentioned above, hesperidin is an abundant bioflavonoid present in all parts of the sweet orange (*Citrus sinensis*) fruit, and can be extracted from these fruits together with other bioflavonoids: neohesperidin, eriocitrin, narirutin, naringin, didymin, sinensetin, hexamethoxyflavone, tangeretin, nobiletin [10]. Extraction of hesperidin and other phenolic compounds from orange peel is very important, as it can convert an agricultural waste product into a valuable commodity. Every year, 100 million tons of different kinds of Citrus are used in juice production, and 40-50% of that mass remains as citrus peel waste [11]. It was estimated that Citrus-Processing Waste from Orange (CPWO) has around 1.5 to 1.8% of hesperidin [9, 12]. Phenol compounds from citrus peel have great potential for functional food, cosmetic, and pharmaceutical applications. Therefore, for their extraction, i.e., for the extraction of hesperidin and other bioflavonoids from orange and other citrus peel waste, different methodologies have been developed.

Hesperidin was first isolated from the white inner layer (albedo) of citrus peel in 1828 by the French chemist Lebreton [13]. Many different procedures have since been described in literature and used for extraction of hesperidin from different plant materials, including maceration [4, 10, 14, 15], Soxhlet extraction [4, 14], ultrasound assisted extraction [10, 16], high hydrostatic pressure extraction [17], microwave assisted extraction [18], enzyme assisted extraction, and supercritical fluid extraction. All of these techniques have advantages and disadvantages when yield of isolated compounds, time and cost of extraction, simplicity of work and environmental effects are compared. In addition to extraction itself, preparation of plant material and purification of isolated compounds are important steps in the isolation of hesperidin and other flavonoids.

The most important steps in the preparation of peel for the extraction are dehydration and milling. Dehydration prevents the degradation of natural compounds and leads to the increase of peel porosity (increase in solvent diffusion) and extraction of compounds. Dehydration is necessary only for supercritical (CO_2) extraction. For the other extraction techniques, fresh or frozen peel material can be used. Dehydration can be performed by air-, oven- or freeze-drying. Different yields of phenolic compounds obtained using fresh or dried peels have been reported. In one case, the yields obtained from dried plant material are higher in comparison to yields from raw fresh material [16], but in others the yields were lower [19]. The differences observed could be caused by different drying conditions of the plant material. If the peel material is dried at high temperatures for too long, phenolic compounds can be destroyed. Milling increases contact surface area between the peel constituents and solvent, and, together with drying, breaks cellular structures and increases the extraction of phenolic compounds. Depending of the type of citrus and extraction technique, after the milling process, particle size of peel should be in the range of 0.2-2.0 mm. For *Citrus sinensis*, a standard particle size of around 0.3 mm was used [10].

Maceration and Soxhlet extractions are considered conventional techniques for the extraction of aromatic compounds from the plant materials. Certain methodological optimizations and use of different solvents (methanol, ethanol, water, alkaline solutions, acetone, and diethyl ether) usually lead to the highest yields of active compounds being obtained when methanol and ethanol are employed [19, 20, 21]. Because of the toxic and hazardous properties of methanol, ethanol is the most appropriate solvent for use in conventional extractions. Extraction with ethanol yielded the highest amounts of hesperidin in comparison to the extraction with methanol and hot water [22]. Optimal concentration of ethanol varies depending of the citrus peel type. For the extraction of flavanones from *Citrus sinensis* peel, ethanol concentration of 80% is proposed as optimal [23]. Optimal time for the extraction depends on applied temperature. When temperature is high, the time of extraction is short, but increase in temperature can lead to degradation of phenol compounds and decrease of their yields.

Novel techniques for the extraction of flavonoids from citrus peel have been developed to achieve higher yields of isolated compounds in shorter extraction times, with low energy and solvent consumption, and with as little as possible degradation of heat labile compounds. One of these techniques is *ultrasound-assisted extraction* with 80% ethanol. This significantly increases yields of hesperidin and other flavonoids from *Citrus sinensis* peel, compared to conventional extraction with 80% ethanol, in mechanical stirrer [23, 24]. This extraction should be performed at a sonication power of 100-150 W, as at higher powers the yields of flavonoids decrease. Ultrasound generates heat, so during the extraction temperature must be controlled (optimal range is 30-40°C). *Microwave-assisted extraction* is performed using power lower than 400 W, with methanol or ethanol as solvents, for a period of between 49 s and 8 min at temperature of 135-140°C [18, 20]. For the majority of flavonoids isolated from orange peel, better yields than with the conventional method were achieved with microwave-assisted extraction at the power of 200 W (corresponding to an extraction at 76°C) [10]. *Supercritical CO_2 extraction* has some advantages (e.g., inexpensive CO_2 and minimal or no use of organic solvent) in comparison with conventional techniques. On the other hand, supercritical CO_2 extraction (during 30 min at 35-80°C and a pressure of 10-22 MPa) gives lower total flavonoids yield from *Citrus sinensis* peel in comparison with conventional solvent extraction. As this technique is more convenient for the extraction of non-polar molecules, the extraction of more polar glycosylated flavanones is less efficient. When this method was used for hesperidin extraction the highest yield was obtained when extraction was performed at 80°C and a pressure of 10 MPa [10]. *High pressure extraction*, due to improvement of enzyme activity and degradation of phenolic compounds in plants, leads to decrease of total phenol contents in comparison to conventional methods. It can also lead to a decrease of the yields of major flavonoids (hesperidin, neohesperidin and naringin) in the extract from orange peel [10].

The method for hesperidin extraction used most frequently in our group is shown to be very efficient. For extracting hesperidin from orange peel, approx. 200 g of biomass is oven dried at 70°C for about 8 h to remove all moisture content, and then powdered in a food processor. 20 g of powdered material is then placed in a thimble made from a thick filter paper and placed inside the extraction tube of a Soxhlet extractor. The extractor is placed onto a 500 mL flask containing 250 mL of petroleum ether as the extraction solvent and magnetic stir bar. The Soxhlet is then equipped with a reflux condenser. The reaction mixture is stirred and heated for 1.5 h, until the petroleum ether extracts all essential oil and the extract in Soxhlet remains colorless. The next step involves methanol (250 mL) extraction and takes 3 h to complete. After that, methanol is evaporated at the rotary evaporator, leaving a syrupy material containing flavonoids glycosides and sugars. This syrup is mixed with 50 mL of 6% acetic acid and left for precipitation. The precipitated yellowy solid is the crude hesperidin. This is centrifuged at 5000 rpm for 5 min. The supernatant is drained off and the hesperidin is washed with a small amount of 6% acetic acid. Crude powder is dried at 60°C until constant weight is achieved.

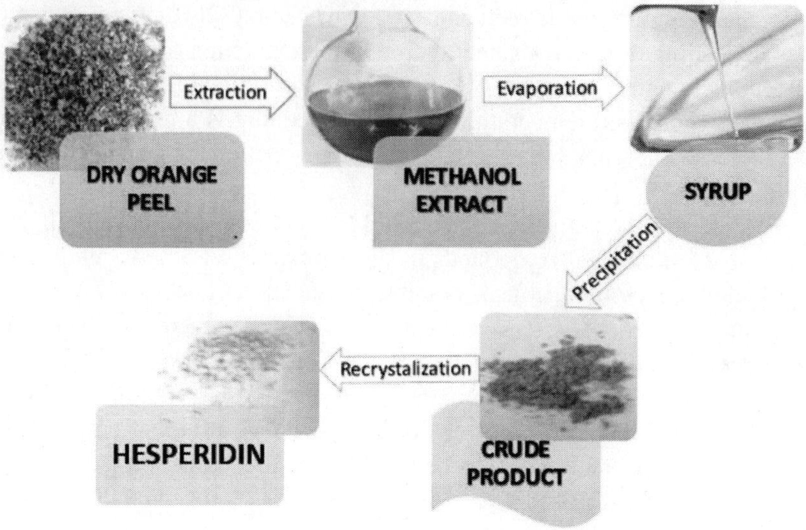

a

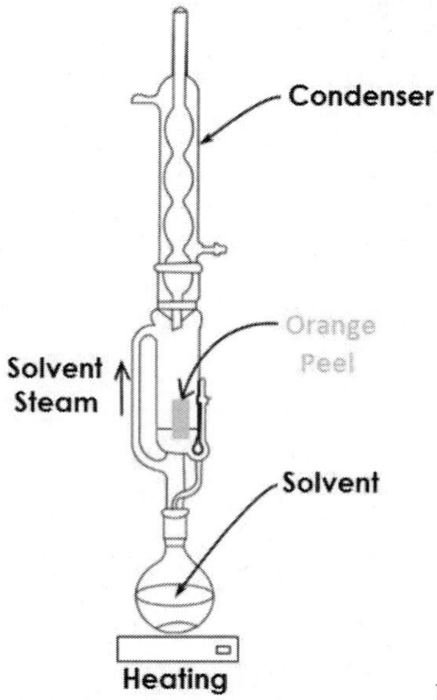

b

Figure 5. Steps in hesperidin extraction from orange peel. Firstly, essential oil is removed using a semi-solid liquid extraction with an apolar solvent. Changing the solvent to a more polar methanol, bioflavones are extracted and concentrated into syrup. Hesperidin is purified using an acetic acid solution (6%) and posterior crystallization from dimethyl sulphoxide (DMSO) with water (a). Extraction is achieved in the Soxhlet extractor equipped with condenser and heater, filled with dry orange peel (b).

For recrystallization of hesperidin, a 5% solution of the crude product in dimethyl-sulfoxide is stirred at 60-80°C. Heating is stopped and the distilled water (at the same amount as dimethyl-sulfoxide) is added dropwise under continuous stirring. The resulting mixture is left for cooling at room temperature and white crystals of hesperidin are precipitated. The mixture is centrifuged at 5000 rpm for 5 min. The hesperidin is washed with a small amount of warm distilled water and then with isopropanol, and dried in desiccator until constant weight is achieved (Figure 5). The yield of hesperidin is about 1.5-2.0%.

5. Spectral Data of Hesperidin

After isolation and purification from orange peel, hesperidin structure (Figure 6) can be determined or confirmed based on NMR and infrared (IR) spectroscopy.

The ^{1}H NMR spectrum of hesperidin (Figure 7, Table 1), shows typical flavanone skeleton signals at δ 5.51 (1H, dd, *J* = 12.1, 3.3 Hz, H-2), three protons of 1,3,4-trisubstitued ring at δ 6.91 (3H, m), two anomeric protons of glucose and rhamnose at δ 4.98 (1H, d, *J* = 7.3 Hz) and 4.53 (H, m), one methoxy signal at δ 3.8 (3H, s), one methyl signal at δ 1.09 (4H, d, *J* = 6.2 Hz), and two aromatic hydroxyl signals at 12.02 (1H, s, 5-OH) and 9.09 (1H, s).

The ^{13}C NMR spectrum also shows characteristic chemical shifts for hesperidin (Figure 8, Table 2).

FTIR spectrum of hesperidin (Figure 9) shown a strong band of OH stretching (str) at 3544 and 3470 cm^{-1}, CH (aliphatic) at 2976, 2916 and 2848 cm^{-1}, C = C (aromatic) at 1606, 1519, 1467 and 1443 cm^{-1} and C = O (str) at 1648 cm^{-1}, C-O (str) at 1298, 1276, 1240, 1203, 1182, 1154, 1131, 1094, 1050, 1033, and 1009 cm^{-1}.

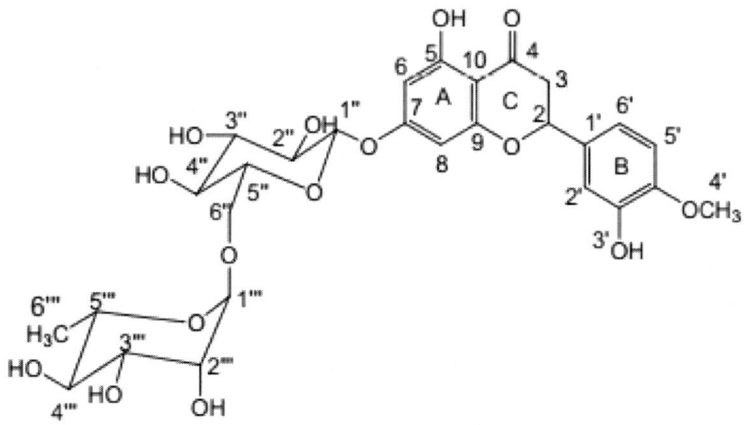

Figure 6. Chemical structure of hesperidin. Numbering of all carbon atoms is given as used for analysis of the NMR data.

Table 1. Hesperidin ^{1}H NMR chemical shifts and assignments

Assignment	Chemical shifts in ppm	Assignment	Chemical shifts in ppm
OH-5	12.0 (s) 1H	H-1'''	4.5 (d) 1H
OH-3'	9.1 (s) 1H	4'-OCH_3	3.8 (s) 3H
H-2',6',5'	6.9 (m) 3H	H-2'''	3.6(m) 1H
H-3'''	3.3 (m) 1H	H-5''	3.5 (m) 1H
H-3	3.2 (d) 1H	H-5'''	3.4 (m) 1H
H-8	6.1 (d) 1H	H-6'', H-3'''	3.4 (m) 3H
H-6	6.1 (d) 1H	H-2'''	3.2 (m) 1H
H-2	5.5 (dd) 1H	H-4'', H-4'''	3.2 (m) 2H
H-1	5.4 (d) 1H	H-6'''	1.1 (d) 3H

Table 2. Hesperidin ^{13}C NMR chemical shifts and assignments

Chemical shifts in ppm	Assignment	Chemical shifts in ppm	Assignment
197.5	C4	96.0	C8
165.6	C7	78.9	C2
162.9	C5	76.7	C3''
162.4	C9	76.0	C5''
148.4	C4'	73.4	C2''
146.9	C3'	72.5	C4'''
131.4	C1'	71.2	C4''
118.4	C6'	70.8	C3'''
114.6	C2'	66.5	C2'''
112.5	C5'	68.8	C5'''
103.8	C10	68.8	C6''
101.1	C1'''	56.2	OCH3
99.9	C1''	42.5	C3
96.9	C6	18.3	C6'''

6. Hydrolysis of Hesperidin

Hesperetin, the aglycone form of hesperidin, has important properties, principally as regards human health [25, 26]. Because the amount of hesperidin in nature is much higher than hesperetin, hydrolysis of the glycosidic moiety of hesperidin is of great importance.

Nowadays, the best way to obtain hesperidin is through extraction from orange peels, with yields of between 1.5 and 1.8% of hesperidin [9, 12, 27]. Once hesperidin has been dried and purified, hydrolysis can be carried out under acidic or enzymatic conditions.

Acid hydrolysis of hesperidin using dilute sulfuric acid can be accomplished in several ways [27, 10, 28].

However, none of the reactions give good results, partly because of the low solubility of hesperidin and partly because of the non-specificity of sulfuric acid, generating several side products [27]. Enzymatic hydrolysis is more specific, and occurs at a wide range of pHs and temperatures. This process can be adapted to the optimal conditions of activity of the enzyme used, and thus only the desired product is obtained.

The hydrolysis of hesperidin by β-D-glucosidase produced by *Pyrococcus furiosus*, with activity of 172 U/g of enzyme, for example, converts all hesperidin to hesperetin in a 10 min reaction conducted at pH 5.5 and 95°C. The addition of 0.5% DMSO solution increases the solubility of hesperidin and favors the reaction [29].

Another example is the use of naringinase, an association of α-L–rhamnosidase with β-D-glucosidase, produced by *Aspergillus sojae*, which produces the hesperetin-7-*O*-glucoside with a yield of 78% [14].

In further studies, using a semi-purified extract of tannin acyl hydrolase (tannase), an enzyme that hydrolyzes esters bonds, produced by *Paecilomyces variotii*, with activity of 1.88 U/mL of reaction, converts the hesperidin to hesperetin in 30 min at pH 7.4 at 40°C, (Scheme 1) [30, 31].

Although the enzymes have different modes of action, once β-D-glucosidase and tannase have detached the rutinose from the hesperetin and naringin, the hesperetin-7-*O*-glucoside and rhamnose are produced. All of these products are commodities of economic interest [10].

Hesperetin, aside from its pharmaceutical uses, can be used in the food industry for the enrichment of food and juices. Hesperetin-7-*O*-glucoside can be converted into an intensely sweet dihydrochalcone by simple hydrogenation in alkaline media. Rhamnose can be used as a sweetener and as a precursor in the synthesis of furaneol (4-hydroxy-2,5-dimethyl-3 (2*H*)-furanone), an important component of synthetic fruit flavors, used in perfumery and as an additive in the food industry [10].

Despite the fact that bio-transformation is an environmentally suitable technique, the cost of commercial enzymes is still too high to work on a large scale and the production of these enzymes is very low.

Therefore, it is also the case that much more research is required in order to determine the optimal conditions of operation and reuse of these enzymes to optimize the process for achieving excellent yields of cited commodities [15].

7. Synthesis of the Bioflavonoids: Shikimate Pathway

The shikimate pathway, in combination with the acetate pathway, is responsible for the production of secondary metabolites in plants and microorganisms, and plays a significant role in the biosynthesis of the flavonoids. Flavonoids are synthesized by plants in response to UV radiation, high or low temperatures, tissue injuries, infection, and effects of reactive oxygen species (ROS). Phosphoenolpyruvate and D-erythrose 4-phosphate are compounds reacted in the first step of the pathway, which lead to the formation of shikimate, precursor to the synthesis of various aromatic compounds. L-phenylalanine, L-tyrosine and L-tryptophan, produced in further reactions, form a basis of phenyl propane (C6-C3) units for cinnamic acid, *p*-aminobenzoic acid, gallic acid, coumarins, lignans, flavonoids, and a wide range of alkaloid structures.

The key intermediates in hesperidin synthesis are presented in Scheme 2, but the chain of the reactions is much more complicated than given. Starter units for acetyl synthesis of the flavonoids are cinnamoyl-CoA and 3 molecules of malonyl-CoA, coupled by enzyme chalcone synthase, followed by a Claisen reaction giving the naringenin-chalcone. Chalcones have the role of being the precursors for a wide range of flavonoid derivatives found throughout the plants.

Most chalcones contain a six-membered heterocyclic ring, formed by Michael-type nucleophilic attack of a phenol group into the unsaturated ketone giving a flavanone. This isomerization can occur chemically, and flavanone and chalcone favor acidic and basic conditions respectively. The reaction is enzyme catalyzed and stereospecific, resulting in the formation of a single flavanone enantiomer in nature [32].

Depending on the plant species, the basic flavonoid skeleton is further modified by enzymes, such as isomerases, reductases, and hydroxylases, leading to the different flavonoid subclasses. Finally, transferases modify the flavonoid backbone with sugars, methyl groups, and/or acyl moieties. Due to these modifications, the physiological activities of the flavonoids, i.e., their solubility, reactivity, and interaction with cellular targets, are modulated [33].

8. Bioactive Properties of Hesperidin and Hesperetin

As already stated, hesperidin and other bioflavones play important roles in citrus plant defenses, but harvest time affects their abundances in the fruit peel. Close connection between fruit growth and hesperidin amounts has been demonstrated [34]. The highest amounts of hesperidin are accumulated in very young tissues inside the fruit, which might indicate a protective function or its importance for the fruit growth and maturation. When consumed by humans, as with, for example, orange and grapefruit fruit or the juices thereof, the main bioflavonoid detected in human plasma is hesperidin. Metabolites of hesperidin and narirutin (such as naringenin-7-O-glucuronide, naringenin-4′-O-glucuronide, naringenin and hesperetin sulfate, then hesperetin-7-O-glucuronide and hesperetin-3′-O-glucuronide), detected in the blood plasma and urine after consumption of fresh squeezed and industrial processed orange juice, indicate different pathways of biotransformation of the orange bioflavonoids [28].

Regarding human health, hesperidin exerts antioxidant, anti-inflammatory, anticancer, antimicrobial, and antiviral properties, leads to a decrease in capillary permeability, and reduces cholesterol and risk of stroke. With the possibility to regulate hepatic cholesterol synthesis by inhibiting the activity of 3-hydroxy-3-methylglutaryl coenzyme A (HMG-CoA) reductase, hesperidin is effectively used as a supplemental agent [28]. A number of studies have described hesperidin's calcium channel blocker activity and improved anti-hypercholesterolemic activity, manifested as decreasing cholesterol LDL, total lipids, and triglyceride levels, and increased HDL levels [35].

8.1. Antioxidant Activity of Hesperidin

Flavonoids are well known for their antioxidant activity, being able to scavenge and inactivate free radicals and reactive oxygen species such as hydroxyl, peroxyl, alkoxyl and peroxynitrite radicals. These species are responsible for ageing mechanisms as well as for inducing many kinds of tissue damage, cancer, atherosclerosis, hypertension, senile dementia, and amyloidosis. Also, free radicals and reactive oxygen species are mostly produced by enzymatic activities related to the maintenance of the organism.

However, environmental factors such as ionizing radiation, exposure to toxicants, and diet can also interfere in the radical network [36].

The radical scavenging property inherent in flavonoids is intimately related to the basic structure of this class of molecules. There are three distinct features that turn flavonoids into excellent antioxidants: the catechol (*o*-dihydroxy) group in the B-ring, the 2, 3 double bond, and the presence of 7-, 5- and 3-hydroxyl groups [37]. Hesperidin possesses only the last feature, which suggests a mild antioxidant activity (Figure 3). Indeed, Garg et al. concluded, based on analysis of several papers in their review work, that hesperidin was inactive or only moderately active in comparison with other flavonoids and nonflavonoidal antioxidants [38]. On the other hand, hesperetin has an additional hydroxyl group in the molecule, which reinforces the fact that it has a stronger antioxidant activity compared to its glycosylated form. The O-glycosylation affects its capacity for electronic delocalization, decreasing the antioxidant activity [39]. Hence, the substitution of rutinose by the hydroxyl group enhances the radical scavenging property of the molecule, and the reason for this lies in the antioxidative mechanism.

In order to study the actual mechanism by which flavonoids (and more specifically, hesperetin) reduce radical and reactive oxygen species, various oxidants, chemicals, and toxins that cause damage to tissues via oxidative stress or other mechanisms have been evaluated [40]. A common approach is the study of the reaction between flavonoids and superoxides or bromine radical species. Kinetic studies, along with potentiometric and acid-base experiments, have enabled a complete elucidation of the oxidation mechanism promoted by flavonoids. Looking at the structure of hesperidin, it is reasonable to affirm that the B-ring is more electron rich than the A-ring because of the contribution of the methoxy group in the *ortho* position. Thus, the first site of oxidation in the molecule is the 3'-phenoxyl group in a one-electron reduction. In fact, organic peroxyl radicals selectively attack the B-ring, forming another radical species (Scheme 3).

The pKa of both phenoxyl groups in the hesperidin molecule are 8.9 and 11.2. However, the pKa of the phenoxyl radical in the 3' position drops to a value between 4 and 5. Given that flavonoids act as antioxidants in the pH range of 7 to 10, it is concluded that the mechanism of oxidation is a concerted redox reaction involving the transference of one electron and one proton. The reduction potential of the phenoxyl radical is strongly affected by substituents in the B-ring.

Hesperidin has an electron-donating group (methoxy) attached to the B-ring, which elevates its reduction potential (E = 0.72 V) when compared to other flavonoids with catechol groups in the B-ring (E = 0.5 – 0.7 V). Nevertheless, the reduction potential is still smaller than those from alkylperoxyl radicals (E = 1.05 V) and superoxide radicals (E = 0.94 V), which confirms that hesperidin is able to scavenge and inactivate these harmful radical species in the organism [41].

In addition to direct radical scavenging, hesperidin and hesperetin also exert their antioxidant properties by augmenting cellular antioxidant defense via the ERK/Nrf2 signaling pathway [42, 43, 44]. Hesperidin activates the Nrf2 and ERK 1/2 pathways, leading to the upregulation of HO-1 expression. This molecule promotes an increase in levels of the antioxidant bilirubin and decreases the concentration of intracellular pro-oxidants [42, 43]. Moreover, HO-1 expression enhances the production of carbon monoxide (CO), which has anti-inflammatory and anti-apoptotic activities in cells. CO also enhances Guanylyl cyclase in cells [41]. Additionally, the levels of antioxidant enzymes, such as SOD, CAT, and GST, are also enhanced by the activation of Nrf2.

The antioxidant properties of hesperidin have been proven by a number of assays [39]. Wilmsen et al. [45] have reported an experiment, which consisted of determining the 1,1-diphenyl-2-picrylhydrazyl (DPPH) radical scavenger capacity of hesperidin *in vitro*. It was found that hesperidin was even more successful in diminishing the concentration of DPPH across the entire concentration range in comparison with trolox, a well-known antioxidant used as the positive control.

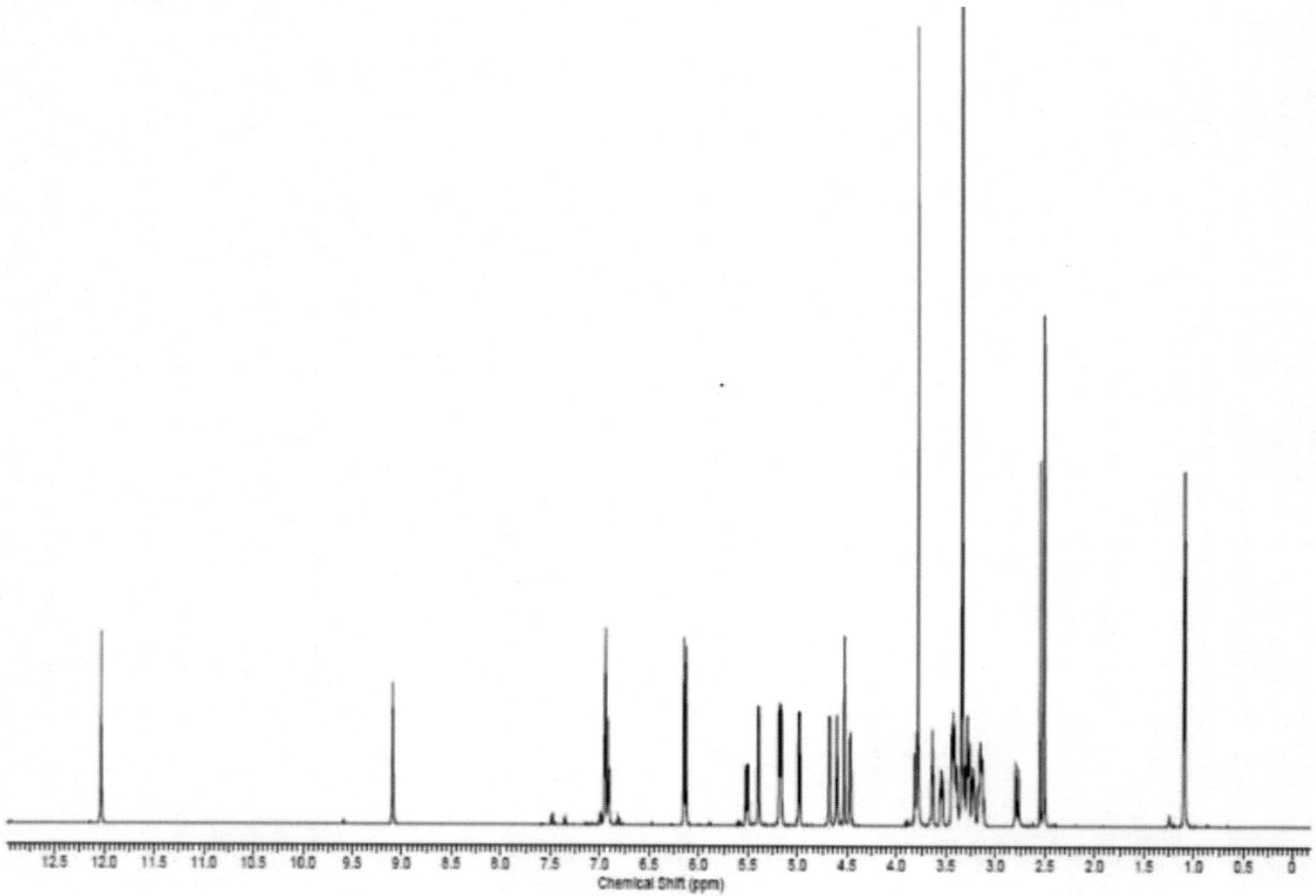

Figure 7. ^{1}H NMR spectrum of hesperidin in DMSO-d_6 as a solvent, on Bruker Advance 600 MHz equipment at 25°C.

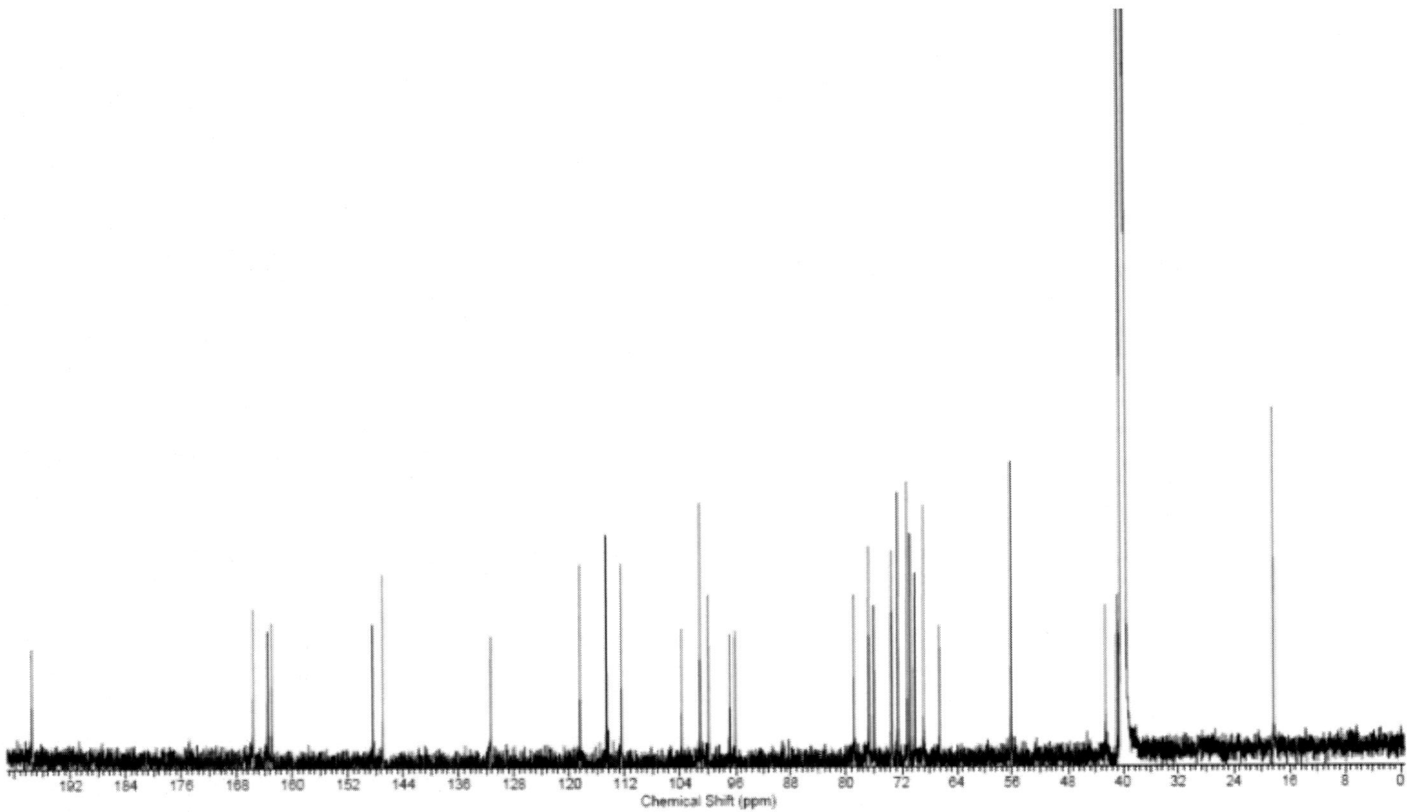

Figure 8. ^{13}C NMR spectrum of hesperidin in DMSO-d_6 as a solvent, on Bruker Advance 600 MHz equipment at 25°C.

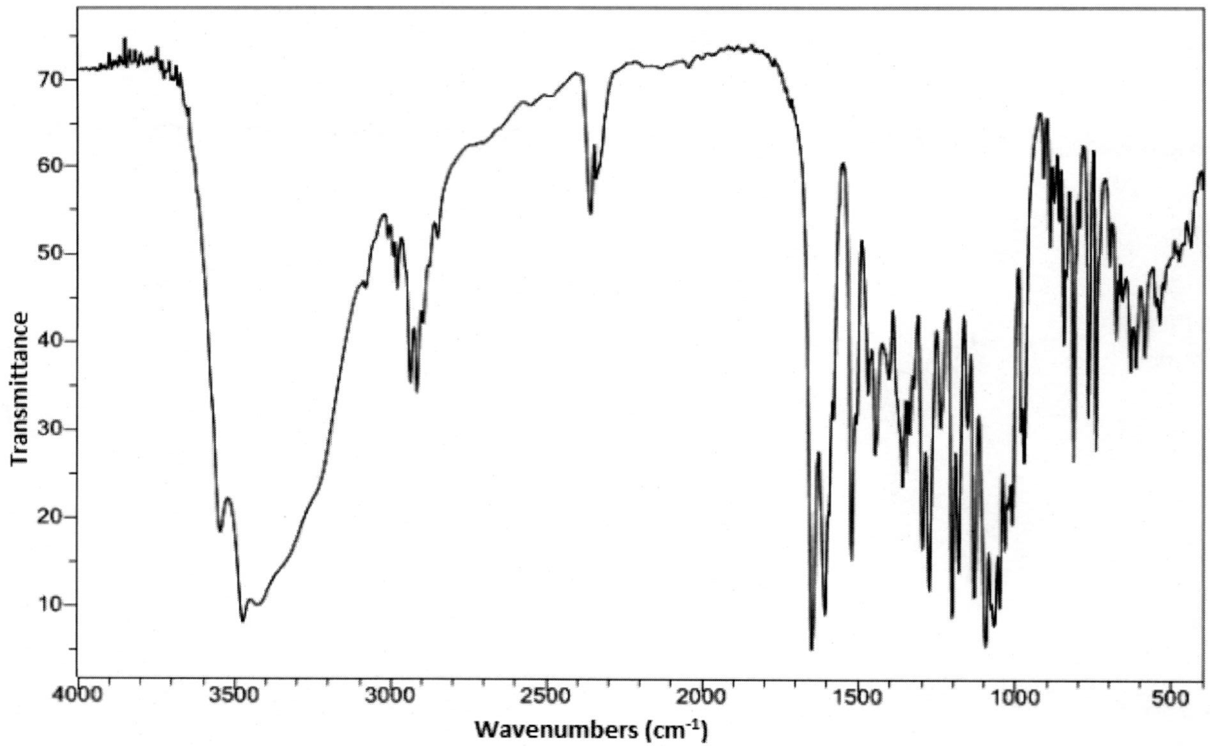

Figure 9. Fourier transform infrared (FTIR) spectrum of hesperidin. 1% of hesperidin in KBr disc placed to Bomem MB 100 (Canada).

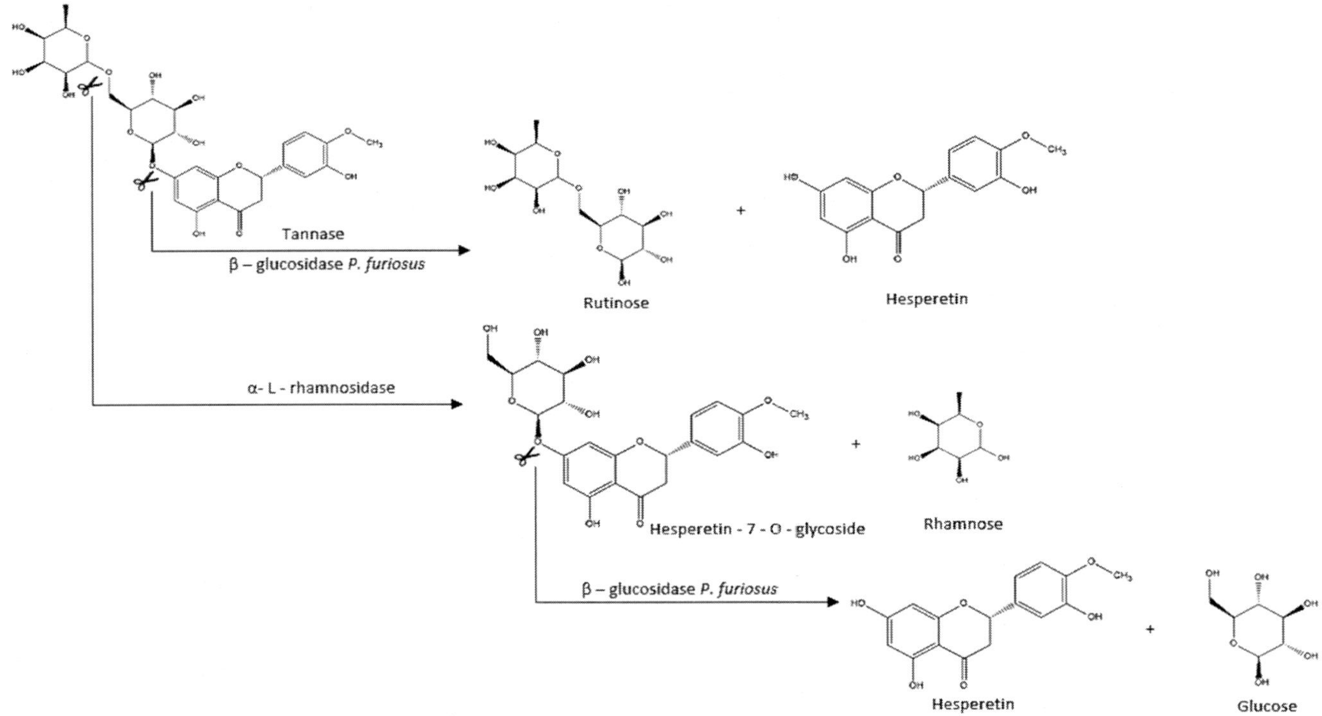

Scheme 1. Selectivity of hydrolases with hesperidin as substrate. The hydrolysis of hesperidin by β-D-glycosidase produced by *Pyrococcus furiosus* converts hesperidin into its aglycone form, hesperetin, in 10 min at pH 5.5 and 95°C. Through the use of naringinase, an association of α-L–rhamnosidase with β-D-glucosidase, produced by *Aspergillus sojae*, the hesperetin-7-O-glucoside is obtained with a yield of 78%, in 30 min at pH 7.4 at 40°C.

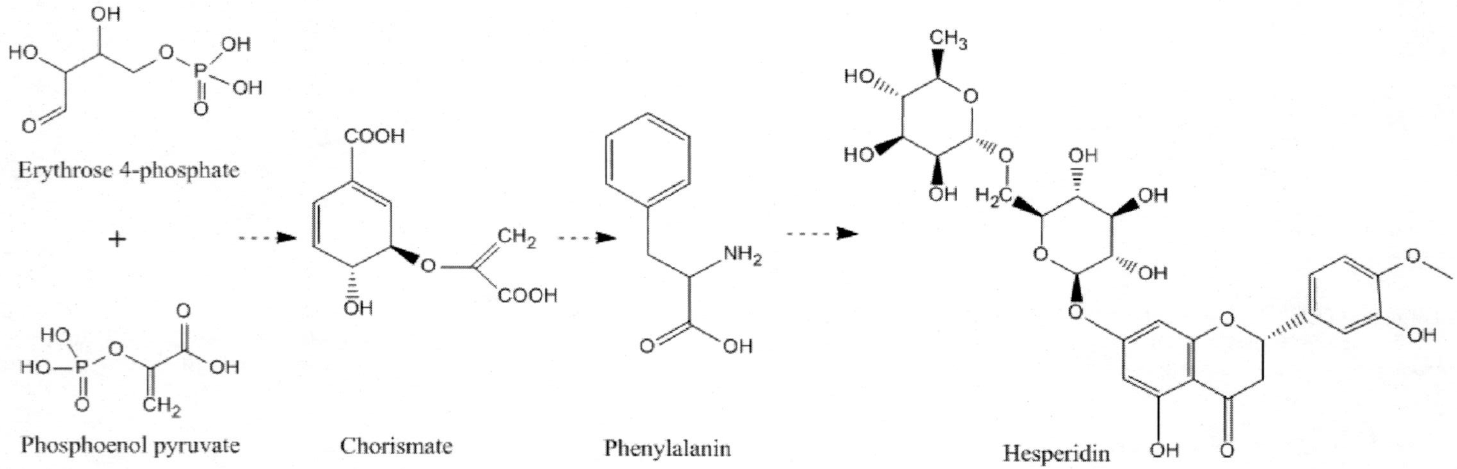

Scheme 2. Hesperidin as a metabolite of the Shikimate pathway, from the biosynthesis of aromatic amino acids in plants, with chorismate and phenylalanine as main precursors of biosynthesis.

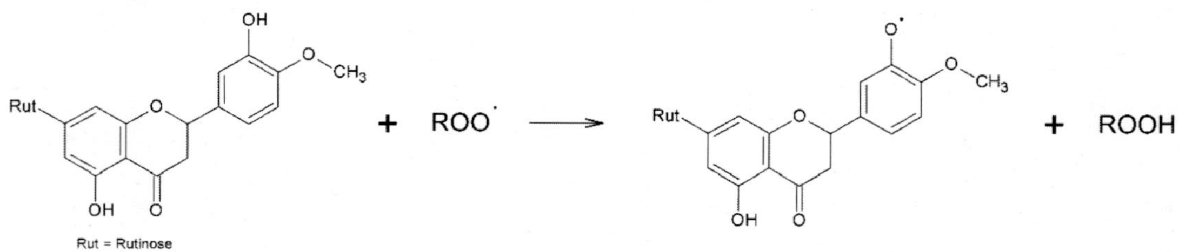

Scheme 3. Proposed antioxidative mechanism involving hesperidin and an organic peroxyl radical.

Our research group has confirmed the antioxidative capacity of hesperidin with an assay involving the formation of silver nanoparticles (AgNP). Silver nanoparticles are easily produced in a solution containing diluted silver nitrate and a mild reducing agent. In this case, hesperidin is used as the reducing agent in order to confirm that it is able to reduce even metal species. Silver nanoparticles are easily detected and characterized by UV-visible spectroscopy as well as with Zeta Potential and DLS (*Dynamic Light Scattering*) experiments. Similarly to the radical species, the Ag^{+}/Ag redox couple has a reduction potential of E = 0.799 V [46], which is higher than the hesperidin phenoxyl radical (E = 0.72 V), showing that hesperidin can indeed reduce silver ions. What we see in practice is that the small gap of reduction potentials between both species is convenient in the reaction; when it comes to silver nanoparticles synthesis, a slow nucleation of silver atoms is required in order to produce uniform *nanoclusters* in the nanometers range.

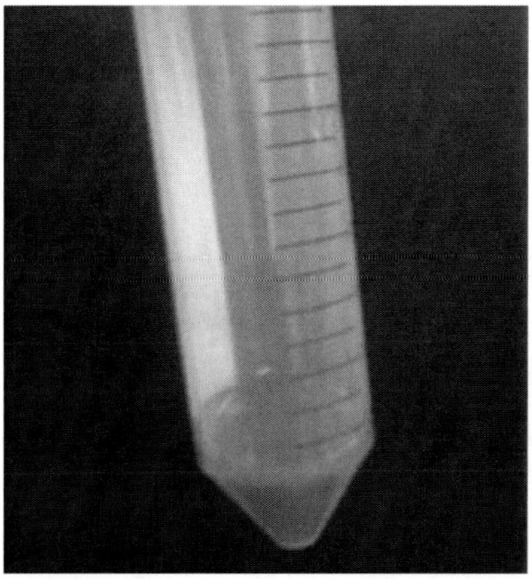

a

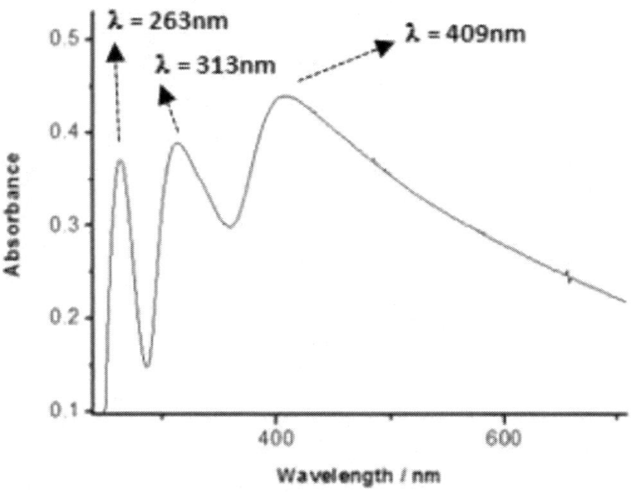

b

Figure 10. AgNP solution synthesized from hesperidin and silver(I) nitrate after 24 h (a) and UV-visible absorption spectrum, showing the surface plasmon resonance band in 409 nm (b). The other two bands in (b) are typical of electronic transitions in flavonoids.

The reaction is performed in alkaline conditions to improve the solubility of the flavonoid. To a flask containing 4.5 mL of a solution of $AgNO_3$ (1.10^{-3} mol L^{-1}) is added 0.5 mL of a hesperidin (0.3 mg mL^{-1}) solution in methanol. A yellowish color quickly appears, denoting the production of silver nanoparticles, which have a surface plasmon resonance absorption band over the range of 400-440 nm, depending on the size of the nanoparticles and the thickness of the capping layer that stabilizes them. The reaction is completed after approximately one hour, yielding nanoparticles approximately 100 nm in diameter, with zeta potential of -45 mV (Figure 10). As AgNP produced by this eco-friendly method are not well stabilized, the nanoparticles aggregate over the course of a few hours, changing the color of the solution from yellow to brownish-yellow.

Our research group was also able to perform the same reaction but with the previous coordination of silver (I) with ammonia, using diluted solution of ammonium hydroxide. The coordination compound $[Ag(NH_3)_2]^+$ is slightly less reactive than the Ag^+ ion itself, seeing as the silver (I) is more electron-rich when coordinated to nitrogen atoms.

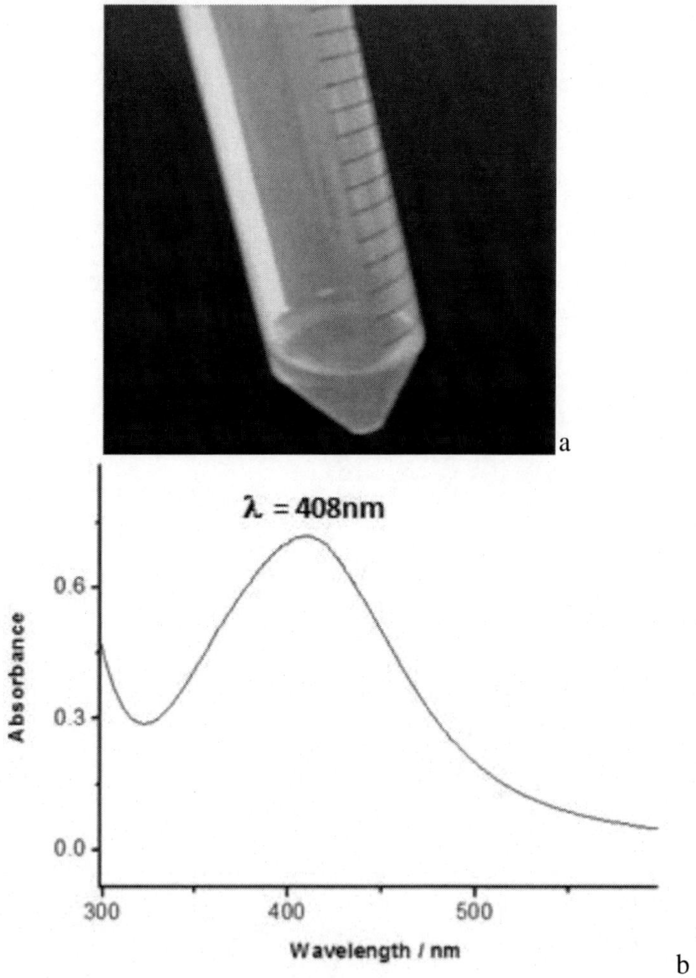

Figure 11. AgNP solution synthesized from a solution of $[Ag(NH_3)_2]^+$ and silver(I) nitrate (a) and UV-visible absorption spectrum showing the surface plasmon resonance band after 1 h (b).

The conditions of reduction are therefore less drastic and steadier, yielding a more stable, small, and uniform set of silver NP [47]. In fact, it is necessary to immerse the solution in a sonication bath for approximately 15 min until the typical yellow color comes into view, with a peak of absorption of surface Plasmon at 408 nm (Figure 11).

Despite the antioxidant capacity of hesperidin, there has to date been no marketing of foodstuffs containing this bioflavonoid in their formulation. Nevertheless, a refreshing drink with vitamin C, hesperidin and extract of citrus flavonoid was recently patented. According to the authors, the components of the drink conduce to a synergistic effect providing antioxidant activity.

Even though they are natural compounds, depending on their chemical structure (hesperidin or hesperetin) those polyphenols may have some side effects on beneficial microbiota by altering micro-ecology in the gut [48]. Thus, more investigation is needed in order to determine the suitable doses and the more appropriate compounds for human consumption.

There are in fact several important factors that impact on the efficiency of citrus flavonoids in clinical studies, including efficacy and safety of hesperidin and its metabolites, bioavailability, appropriate dose, patient population, and length of study [49].

8.2. Chemoprotective Potential of Hesperidin and Hesperetin against Carcinogenesis

Despite the low bioavailability (3-6%) and poor permeability ($1x10^{-6}$cm/s) of hesperidin in humans [50], some studies have demonstrated a chemoprotective activity of this bioflavonoid and its aglycone (hesperetin) against carcinoma cell proliferation as well as cell damage with mammalian models. The first findings on the antiproliferative effects over cancer cells by hesperidin were observed in different tissues of cancer-induced in rat. The inhibition effect of hesperidin (1000 ppm) in azoxymethane-induced colon cancer was examined during the initial phase (adenoma) and by the incidence and multiplicity in the final phase after 32 weeks (adenocarcinoma). In addition, the frequency of the aberrant crypt foci (ACF) was analyzed, a biomarker of precursor lesions caused by external chemical factors in the colon of rodents and humans, as well as the ornithine carboxylase activity in this section of the large intestine, and also the blood polyamine level, to confirm the decrease of colonic cell proliferation [51].

Similar procedures were performed for 4-nitroquinoline 1-oxide-induced tongue cancer, resulting in a reduction in the frequency of 75% of the tumors (oral neoplasms) for hesperidin in the initial phase and a reduction of 62% after one week of the induced cancer. A combination of 900 ppm diosmin and 100 ppm of hesperidin (drug Daflon®) showed a reduction in incidences of 69% in the initial phase and 77% after one week [52].

Through the results obtained with MCF-7 human breast cancer cells it was possible to define the IC_{50} (half maximal inhibitory concentration) for various flavonoids. Among them, galangin was the most effective, with IC_{50} = 4.2 μg.mL^{-1}; naringenin was the least effective, with IC_{50} = 18 μg.mL^{-1}, and hesperetin with an IC_{50} concentration of 12 μg.mL^{-1} was effective in inhibition of the cancer cell growth for 8 days [53].

The antineoplastic activity of hesperetin has also been studied with B16-F10 melanoma cells for *in vivo* testing with 10 μmol.L^{-1} concentration for 24, 48 and 72 h.

In vitro transglutaminase activity (TGase, EC 2.3.2.13) has also been monitored, along with the concentration of the products thereof (putrescine, spermidine and spermine), once they were established as factors associated with induction of cancer cell differentiation during the inhibition of carcinogenesis. An increase in TGase activity and reduction in cancer cell proliferation were observed by oral administration of naringenin and hesperetin in mice [54]. Further, through a computerized image analyzer it was demonstrated that the metastases reduction, when compared with untreated animals, is around 69% for naringenin-treated mice and around 36% for hesperetin-treated mice in these conditions [54].

The inhibition of carcinogenesis by hesperetin in human prostate cancer PC-3 cell line was evidenced by Sambantham et al. in 2013 [55].

The results obtained with *in vivo* assays exhibited IC_{50} values of 90 and 40 $\mu mol.L^{-1}$ for 24 and 48 h, respectively, while an increase in the expression of pro-apoptotic gene BAX as well as an attenuation of the BCL_{XL} mRNA gene expression was observed by real-time PCR analysis. The effects of the endogenous hesperetin, a decrease in the BCL_{XL} level, an antiapoptotic protein, and the increase of the protein BAX level continue to cause the reduction of the carcinoma cells' proliferation by the release of cytochrome-C from the mitochondrial membrane, ultimately activating the caspase cascade-mediated apoptosis.

In addition, the study relates a possible mechanism for these effects based on *in silico* analysis. An intermolecular binding of the hesperetin to the DNA binding domain NF-kB is thus associated with a blocking action over this domain, a protein complex that induces antiapoptotic gene expression in response to external factors such as stress, free radicals, ionizing radiation, ultraviolet radiation, and carcinogens, among others [55].

The inhibitory effect of hesperetin and its lactone in colon cancer and their apoptotic mechanisms were also associated with an increase in the level of protein BAX in HT-29 human colon adenocarcinoma cell line. The IC_{50} values for hesperetin lactone were lower than for hesperetin, and higher concentrations were required for a 24 h treatment period (70 $\mu mol.L^{-1}$ for hesperetin and 32 $\mu mol.L^{-1}$ for hesperetin lactone), compared to treatment periods of 48 h (42 $\mu mol.L^{-1}$ for hesperetin and 27 $\mu mol.L^{-1}$ for hesperetin lactone) and 72 h (31 $\mu mol.L^{-1}$ for hesperetin and 21 $\mu mol.L^{-1}$ for hesperetin lactone). In addition, the study shown a decrease in the level of BCL_{-2}, an antiapoptotic protein that inhibits the lipid peroxidation, thus blocking the cytochrome C release from the mitochondrial membrane [56].

The apoptotic mechanism for the inhibition of hepatocellular carcinoma (HepG-2 cells) and gastric cancer cells was examined via *in vivo* and *in vitro* assays. The hesperetin-induced apoptosis was associated with intracellular reactive oxygen species (ROS), whose level was enhanced with the increase of each dose of hesperetin with H_2O_2 pretreatment (1 $mmol.L^{-1}$). The effect of this pathway is the enhanced level of Ca^{2+}, which promotes the ATP (adenosine triphosphate) synthesis, thus inducing oxidative phosphorylation, in order to modulate the mitochondrial permeability transition pore (mPTP) [57].

A global mechanism of cancer prevention by dietary hesperidin and its aglycone form (hesperetin) is still unclear. However, while tests *in vivo* are needed to confirm the mechanism involved in hesperidin's action as blocking agent and radioprotective agent, the studies regarding its therapeutic effect have been conclusive.

8.3. Vessel Protection Properties

The beneficial effects of the oral and supplement use of hesperidin and its aglycon on the cardiovascular system include cardioprotective effects against oxidative stress and ischemia, hypolipidemia, and hypoglycemia. They also affect hypertension, coagulation, and platelet aggregation, and can lead to the reduction or amelioration of cardiovascular diseases including myocardial infarction [58].

The results of recent clinical studies of the beneficial effect of hesperidin, hesperetin and their derivatives in the treatment of heart failure and cardiac remodeling mark them out as good candidates for treatment of various cardiovascular disorders. The condition and proper functioning of endothelial cells have a significant role in all cardiovascular disorders [58].

8.3.1. Effects on Endothelial Cell Function

The integrity of vascular vessels is controlled by endothelial function, and any alteration can lead to the initiation of atherosclerosis and hypertension, the first signs of cardiovascular risk factor. The vascular tone of endothelial cells is dependent on the balance between vasorelaxing compounds, such as nitric oxide (NO), and vasoconstructive substances, such as endothelin-1. NO is synthesized by endothelial nitric oxide synthase (eNOS), and at the same time inactivated by reactive oxygen species generated by vascular NAD(P)H-oxidase. The balance between these processes determines NO bioavailability in the endothelium. Experimentally, hesperetin has been shown to increase both NO production and eNOS expression or activity of endothelial cells by stimulating the phosphorylation of the mediators of NO production in endothelial cells, Akt, AMP kinase, and eNOS. Simultaneously, mediated by estrogen receptor α, hesperetin may have an influence on the NO production, while *in vivo* results have demonstrated the ability of hesperidin to reduce mRNA expression of aortic NAD(P)H oxidase and reduce the NO inactivation. *In vitro* studies confirm that among all flavanones, hesperetin may improve endothelium function by increasing the availability of NO [58, 59, 60].

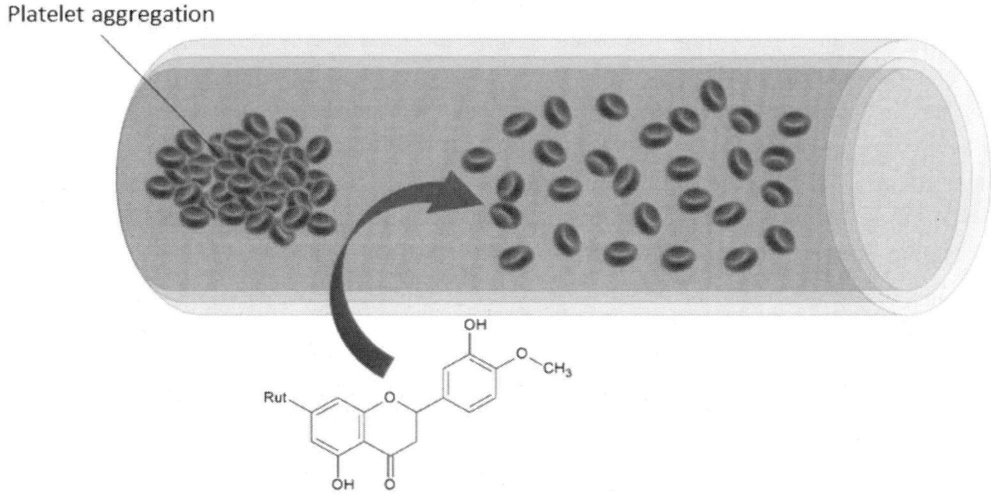

Scheme 4. Entrance of the hesperidin into the epithelial cell acting as anti-aggregation agent.

8.3.2. Effects on Platelet Aggregation

Hesperidin has been reported to inhibit epinephrine- and ADP-induced platelet aggregation, and also to reduce aggregation of erythrocytes (Scheme 4). Hesperetin, via inhibition of arachidonic acid, which is involved in induction of serotonin secretion, can have an antiplatelet function [61].

This effect occurs via the inhibition of PLC-γ2 phosphorylation and cyclooxygenase-1 (COX-1), due to serotonin effective participation in platelet aggregation [58]. Combination of hesperidin and diosmin decreases leukocyte adhesion to the endothelium in postcapillary veins, after ischemia/reperfusion, as well as significantly decreasing ADP-induced platelet aggregation and increasing platelet disaggregation, in rats. Binding of fibrinogen to ADP-induced platelets aggregates was also reduced significantly [38].

8.4. Anti-Inflammatory Properties of Hesperidin

Basically, the inflammatory process begins with tissue damage, which can occur as a result of diverse factors, such as infection, ischemia, antigen-antibody interactions, chemical irritation, thermal action, or mechanical injury. The injury leads to the release of histamine into the bloodstream, promoting an increase in capillary permeability and decrease in blood pressure in the injured site. The blood spreads into the interstitial cell space and takes a longer time to return to the capillary [62, 63, 64]. Thus, a larger amount of blood reaches the injury, leaving the location reddish, warm, and swollen.

In addition to histamine, at the site of inflammation chemostatic mediators are produced, and these attract neutrophils and macrophages to the area. These cells can perform phagocytosis of the factors that led to inflammation, generating other inflammation factors, such as cytokines, chemokines, bradykinins, prostaglandins, and leukotrienes. Simultaneously, platelets and blood coagulation factors also accumulate at the site to prevent possible bleeding. Moreover, endothelial cells produce adhesion factors to perform the diapedesis of monocytes and other circulating inflammatory cells [28, 52].

Pain is a side effect of the process, occurring as the stimulation of nerve endings in the inflammation site, mainly due to prostaglandins and bradykinins and partly by the pressure caused by tissue swelling [53].

The inflammatory process has a complex mechanism that produces various pro-inflammatory mediators through the activation of different signaling pathways [40]. It may occur in three different phases. The acute phase is given by vasodilation and increased capillary permeability, while in the subacute phase there is chemotaxis and diapedesis, and in the chronic phase the degeneration of the tissue and the occurrence of fibrosis mediated by prostaglandins occur [52].

There is evidence that flavonoids can act in all stages of the inflammatory process, but in each one with a specificity and intensity. The quercetin and naringenin, for example, are able to act at all stages of inflammation, while the hesperetin preferentially acts in the acute and chronic phases. In the acute phase of inflammation hesperidin acts to decrease the swelling, having an antagonistic effect on the histamine, decreasing the permeability of blood capillaries. A drug containing a hesperidin and diosmin association is currently marketed under the name Daflon®, primarily used to reduce varicose veins and swelling in the legs [51]. In a synergistic action of several flavonoids, association results are the best [65].

In chronic inflammations, for example, there is an increase of mediators such as inducible nitric oxide synthase (iNOS) and cyclooxygenase – 2 (COX-2), responsible for increasing the levels of nitric oxide and prostaglandins, respectively [28].

In studies performed using the LPS-induced inflammation model, using macrophages that produce high levels of pro-inflammatory cytokines, it was observed that hesperidin decreases levels of nitric oxide by the suppression of iNOS proteins [28, 53, 54], while COX-2 level was kept constant. However, using a mixture of nobiletin, naringin and hesperidin, a decrease was observed in levels of both nitric oxide and COX-2 [28].

The suppression of the expression level of COX-2 is important, since this enzyme is responsible for the regulation of production of prostaglandins. In chronic inflammation, it alters the expression of metalloproteases and promotes the generation of new blood capillaries, increasing the inflammatory process. For this reason, COX-2 is a target of pharmacological action of non-steroidal anti-inflammatory drugs [51].

Combatting inflammation is of great importance, because eliminating inflammation is the best way to prevent chronic diseases such as cancer, neurological diseases, metabolic disorders, and cardiovascular disease [28]. Thus, there remains a great deal of intensive research being conducted in the discovery of new anti-inflammatory drugs, today focusing on natural products and derivatives [28, 54].

9. Oral Uses and Bioavailability of Hesperidin

Polyphenols have a different bioavailability, as well as a minimum amount that should be consumed to disclose their role in our metabolism or target tissue. Among them, gallic acid and isoflavones are the ones that exhibit greater absorption, followed by catechins, flavanones, and quercetin, due to different absorption kinetics [34]. In this chapter, the polyphenols of interest are the flavanones, a flavonoid subgroup basically represented by naringin and hesperidin [55].

Firstly, in order to evaluate the oral intake and absorption of flavonoids by the organism, the form in which flavonoids are used (juices, extracts or pills) is important. In addition, it is crucial to know what other foods are consumed, since a number of studies have shown that the individual's diet may cause changes in the absorption and bioavailability of flavonoids. As an example, when strawberries are consumed with cream, anthocyanin present in the strawberry has a slower absorption and stays longer in the body before being excreted in the urine [66, 67, 68].

In regard to juice intake, the bioavailability of flavonoids is related to their solubility. Since they are slightly soluble in water, they can either be added to macromolecules such as pectin and certain proteins, or simply precipitate, forming part of the cloud of sediment formed during the manufacture of juices [56].

Another important factor is the thermal treatment the juice has undergone during processing to increase its shelf life. This causes degradation of flavonoids, in addition to change in flavor and loss of vitamin C. New technologies have been developed using high pressure and pulsed electric field, requiring a smaller thermal treatment and thus preserving the juice quality. Moreover, there is a decrease in the particle size, which ends up increasing the solubility [55, 56, 61].

Therefore, in a second approach it is necessary to be aware of how and where the flavanones are absorbed and metabolized in the body. Flavanones, aglycone and anthocyanins can be absorbed in the stomach by bilitranslocase, an organic anion transporter expressed in the liver, kidney, vascular endothelium, and gastric epithelium, responsible for the transport of bilirubin and biliverdin. Flavanones aglycone can also be absorbed by the small intestine, getting into the blood plasma shortly after being consumed [5].

Flavonoid glycosides go through the stomach unchanged and are absorbed in the intestine. Although some flavonoids are absorbed by the small intestine, flavanones with a sugar moiety are mostly absorbed later, taking from 5 to 7 h to reach maximum concentration (C_{max}) in the plasma. However, there is still a lot of discussion on this, as, for example, flavones ingested in juices, pills or extracts can arrive faster than when ingested with food. This aspect comes up as a complicating factor to the analysis [69, 70].

The form most conducive to flavonoids' absorption is the aglycone one. However, it is rare to find this form in food that usually contains glycosides [5, 55]. The hydrolysis of sugar moiety, which in the case of hesperidin is formed by a glucose molecule attached to a rhamnose molecule, is performed by α –L-rhamnosidase (EC 3.2.1.40) and β – D-glucosidase or endo–β-D–glucosidase (EC 3.2.1.21) enzymes secreted into the lumen of the intestine by intestinal microflora. These enzymes can transform hesperidin to the aglycone or hesperetin-7-*O*-glucoside, depending on the enzymes that acted [60, 61]. This step is critical for an increase in the bioavailability of flavanones in the body, but it is dependent on the microbiome, gender, and age of each individual [60, 61].

Due to high lipophilicity, the aglycone forms of flavanones are able to cross the barrier of the intestinal epithelial cells by passive diffusion. However, this can also be a limiting factor for their bioavailability, because these cells contain an efflux system formed by proteins containing an ATP-binding cassette (ABC) that can both return the metabolites for gut lumen and allow entry of such compounds into the bloodstream. The absorption of the glycosidic form, hesperetin-7-*O*-glucoside, which contains only one glucose molecule, is only possible by a sugar transporter such as sodium-glucose co-transporter (SGLT1) [5, 62].

The proteins containing the ABC domain may also be responsible for transporting glucuronidated and/or sulfated flavonoids from the small intestine to the inside of the enterocytes [5]. Sometimes, the aglycon moiety can also undergo degradation by the microorganisms to hydroxyl-phenyl propionic acid and/or benzoic acid derivatives.

The presence of these compounds is found during *in vitro* assay, as well as in the plasma and urine of some volunteers, being a process that adversely affects hesperidin and hesperetin bioavailability [55, 56, 60, 71].

9.1. Biotransformation of Hesperidin

Once inside the enterocytes, the aglycone forms of flavanones undergo coupling reactions, known as phase II metabolism reactions, where mainly glucuronidated and hesperetin-7-*O*-glucoside has its glucose removed. Polyphenols and their metabolites are then transported to the liver where are sulfated and glucuronidated [5, 56, 62]. These reactions occur with any non-polar xenobiotics, which are absorbed by our organism. As well as glucuronyl and sulfate groups, the methyl group is also widely used. The consequence of these reactions is the formation of more hydrophilic compounds, which can be excreted as xenobiotics by the kidneys through the urine [5, 62].

In the case of flavanones, the main and most important reaction is the so-called glucuronidation performed by UDP-glucuronosyl transferases (UGTs), followed by the reaction of sulfation catalyzed by sulfotransferases (SULTs). However, the glucuronidation and sulfation mechanisms are not well elucidated by the existing 22 UGTs and ten different SULTs, due to the amount of enzymes that can participate in these processes. However, there is evidence that only the UGT1A1 and UGT1A9 enzymes are involved in the glucuronidation, while sulfation is catalyzed by the SULT1A1, SULT1A2, and SULT1E1 in C3 position, and the SULT1C4 and SULT2A1 in C7 position, the only positions that can enable both sulfation and glucuronidation of flavanones [62].

After this step, the metabolites leave the liver and are transported by the bloodstream to the target tissues and the kidneys, where they will be excreted. In plasma and urine they appear mainly as hesperetin-7-*O*-glucuronide or hesperetin-7-*O*-sulfate, followed by hesperetin-3-*O*-glucuronide or hesperetin-3-*O*-sulfate. This may also occur in the form of hesperetin sulfoglucuronide and hesperetin diglucuronide in smaller quantities (Figure 12). These have also been observed for naringenin, the aglycone form of naringin [55, 57, 60, 62].

Some glucuronide and sulfate flavonoids can be eliminated by the bile, reaching the gallbladder through bilitranslocase present in the liver and subsequently the intestine.

A certain proportion of flavanone can be reabsorbed by the intestine (after the removal of glucuronic moiety by the glucuronidase of microorganisms), increasing the residence time of flavonoids in the body, and the rest is eliminated through feces (Scheme 5) [25].

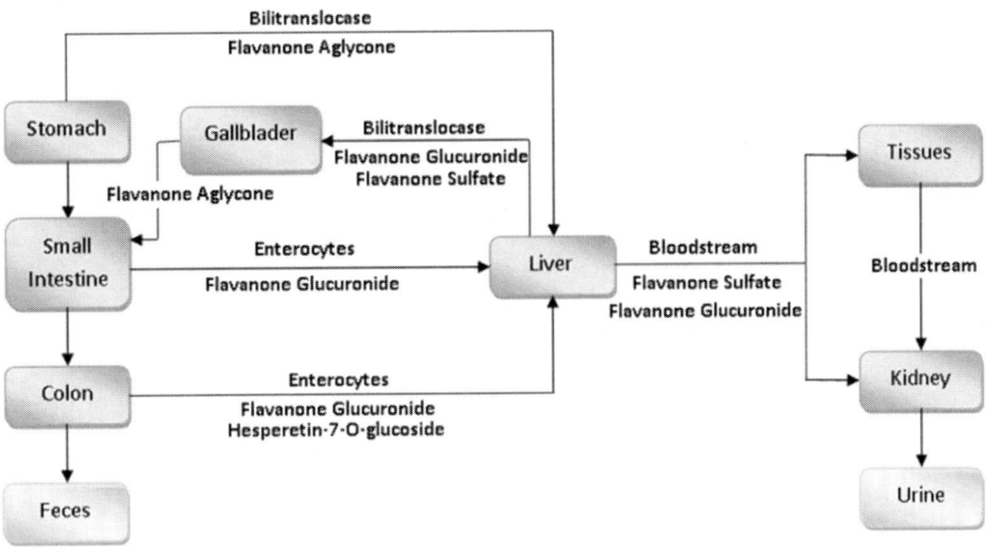

Scheme 5. Flavanone absorption and distribution of metabolites in the human body. The aglycone form is absorbed by the bilitranslocase enzyme in the stomach or the small intestine, while the glycoside form is hydrolyzed by micro-flora and absorbed by the colon. The hesperetin-7-*O*-glucoside is absorbed by a sugar transporter such as sodium-glucose co-transporter (SGLT1). Inside the enterocytes the hesperetin (aglycone form) is glucuronidated and the hesperetin-7-*O*-glucoside has its glucose removed. The hesperetin and the metabolites are transported by the portal vein to the liver where they are sulfated and/or glucuronidated. After this step, the metabolites leave the liver and are transported by the bloodstream to the tissues and the kidneys, where they will be excreted in urine or can also be eliminated by the bile, reabsorbed by the small intestine, or eliminated through the feces [9, 64, 69].

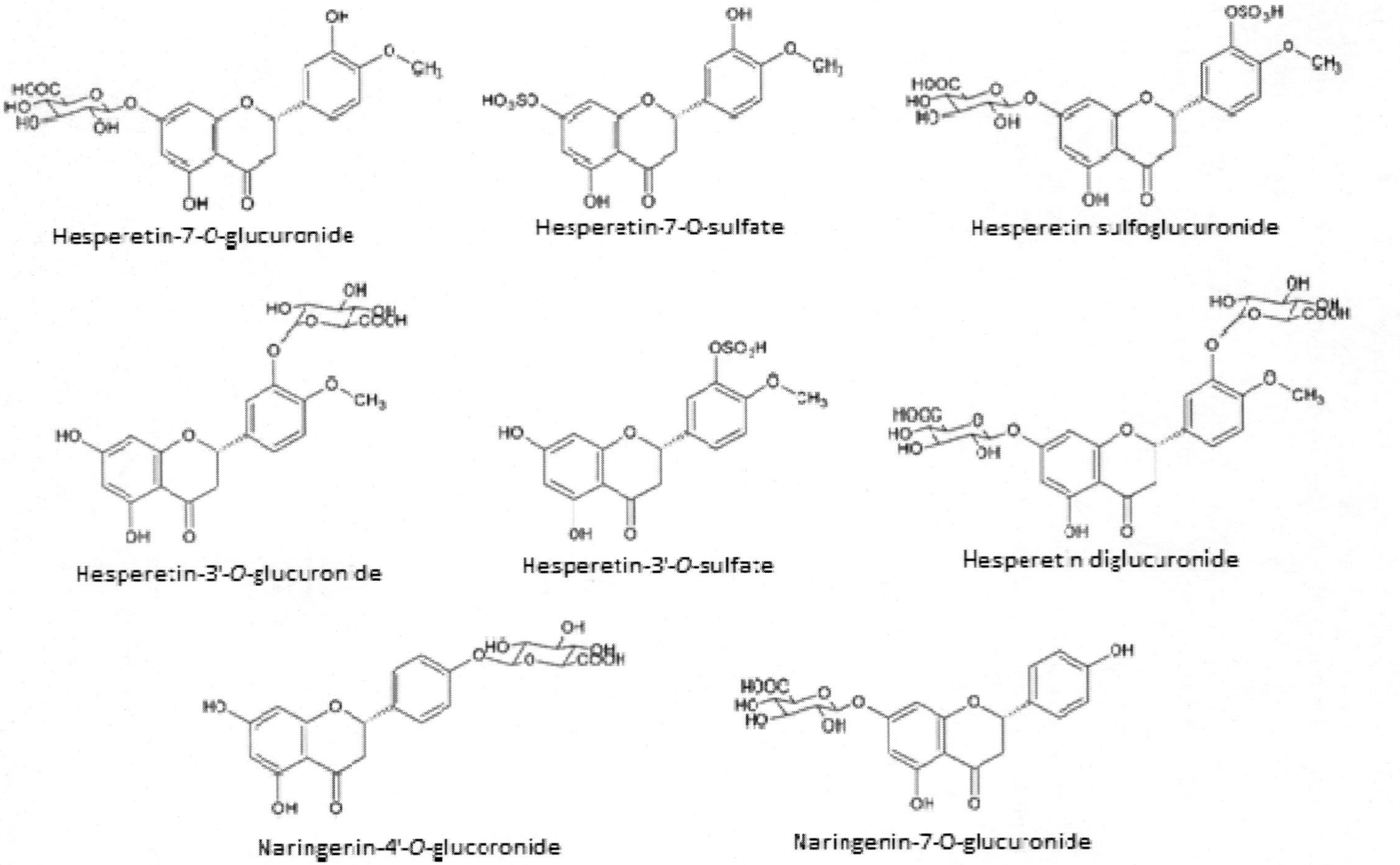

Figure 12. Chemical structures of the hesperetin and naringenin metabolites formed inside the enterocyte and/or liver after oral consumption of these flavones by oral intake of orange juice.

9.2. Improvement in Bioavailability of Polyphenols

Polyphenols, although having many interesting properties *in vitro*, are not able to display the same properties as efficiently *in vivo*, due to several factors, which reduce their bioavailability.

In a normal diet, polyphenol blood concentration is about 1 nmol.L^{-1}, and it rises to 1 μmol.L^{-1} in a diet rich in polyphenols.

However, the ideal concentration should be greater than 1 mmol.L^{-1} since this is the threshold from which the polyphenols show their interesting properties [5].

Various methods have been employed to increase the bioavailability of polyphenols, such as subcutaneous injection of polyphenols in the aglycone form, reduction of the particle size down to micro or nanoscale, encapsulation of the compound in microspheres, nanoparticles, dendrimers, or cyclodextrins and, finally, the use of enzymes aiming at a higher solubility and absorption of these molecules [5, 25]. However, subcutaneous injection is a less suitable method, because, depending on the concentration and the characteristics of the polyphenols, nephrotoxicity may occur in a continuous exposure, as in the case of quercetin applied in doses from 60 to 1700 mg m^{-2} of body surface area [5].

On the other hand, the addition of enzymes into the orange juice may increase the bioavailability of flavanones present in the juice. This promotes hydrolysis of the glycosylated flavanones by enzymes to their aglycone forms [9, 72]. These flavanonc aglycones have higher absorption, because they are absorbed faster and in a wider range of the digestive system [5].

The use of tannase (an enzyme responsible for the hydrolysis of esters bonds in tannins and β-glycosidic bond between glucose and flavanone) as an additive in orange juice is extremely effective in enhancing the antioxidant activity of the juice, potentially increasing this activity by 50-70%. The use of tannase also has the advantage of reducing the bitterness due to the hydrolysis of flavanones linked to neohesperidose [15, 16].

In another study, volunteers drank orange juice without modification, juice containing hesperidinase, and juice with the addition of hesperidin. The volunteers who drank the juice containing hesperidinase, an association of α-L-rhamnosidase with β-D-glucosidase, shown a concentration of metabolites (from hesperetin and naringenin) in plasma four times higher than those who consumed the juice without addition of the enzyme. Those who drank the juice with added hesperidin had shown an increase in the concentration of these metabolites only 1.5 times higher than those who consumed the unmodified juice [65]. Although hesperidinase also produces hesperetin-7-*O*-glucoside, the concentration of metabolites in the plasma of these volunteers was still significantly higher than in those who consumed the beverage without addition of enzymes, since the hesperetin-7-*O*-glucoside is easily transported by a sugar transporter such as sodium-glucose cotransporter (SGLT1) into the enterocytes. There, a glucose molecule is detached from the hesperetin molecule by a β-D-glucosidase present in the cytosol of these cells [4, 39, 40, 42].

Another way to increase the bioavailability of polyphenols could be through the use of cyclodextrins, a polysaccharide that consists of glucose molecules linked by α-1,4 bond forming a frustoconical structure with a polar surface and a non-polar cavity. Cyclodextrins are widely used to encapsulate non-polar molecules, mostly drugs, in order to increase their solubility in an aqueous medium, such as blood. They form a guest-host complex, enhancing the bioavailability of the host [73, 74].

Cyclodextrins are classified in three groups: α, β and γ, constituted of six, seven and eight glucose molecules respectively. The most suitable cyclodextrins to encapsulate polyphenols, in particular flavonoids, are the β-cyclodextrins. The hydroxypropyl-β-cyclodextrin (HP-β-CD) is the most appropriate one, due to its greater solubility and lower toxicity [75, 76]. Studies performed *in vitro* have proven that a complex formed between HP-β-CD and naringenin was able to cross the Caco-2 cells (which simulates the human gut monolayer of epithelium cells) and was absorbed 11 times more than the uncomplexed naringenin. The same study performed with rutin as the host molecule resulted in a 10 times higher absorption of the complexed form when compared to the free form [77, 78].

Furthermore, in an experiment performed *in vivo* using rats, it was demonstrated that the intake of the HP-β-CD/naringenin complex increases in 14.6 times the maximum concentration of naringenin in the blood plasma in comparison with the intake of the flavanone by itself. This is evidence that, in both cases (*in vitro* and *in vivo*), HP-β-CD can cross the barrier of epithelial cells, and that, with the increase of the solubility of naringenin, its absorption is also enhanced, favoring its bioavailability [67, 68].

9.3. Marketed Drugs Containing Hesperidin

Even though hesperidin has several interesting pharmaceutical properties, it is not marketed as a specific drug (medicine), but as an association with other drugs, as in the case of Daflon®, which is composed of 450 mg of diosmin and 50 mg of hesperidin, and in the case of Dactil-ob®, composed of 100 mg piperidolate hydrochloride, 50 mg hesperidin and 50 mg ascorbic acid. Daflon® is a medicine used for chronic venous insufficiency of the lower limbs, varicose veins, swelling and heaviness in the legs, varicose ulcers, and hemorrhoids.

The diosmin acts against venous disorders and increases peripheral circulation return, while the hesperidin is responsible for increasing the speed of blood-flow in the veins by making the blood more fluid and by enhancing the capillaries' resistance, making the microcirculation easier and increasing lymphatic drainage.

Dactil-ob® is a medicine used to prevent premature births. Piperidolate hydrochloride, its main component, reduces muscle spasms, preventing uterine contractions before the correct time of gestation, and is widely used in high-risk pregnancies. Vitamin C, along with hesperidin, is responsible for the integrity of the blood capillaries and avoiding possible bleeding, because hesperidin has vasoprotective action and vitamin C maintains the proper blood levels, helping to protect the fetus [79].

9.4. Regulatory Drugs Containing the Flavonoid

The institution responsible for the regulation and supervision of medicinal products in Brazil is the National Agency of Sanitary Surveillance (Agência Nacional de Vigilância Sanitária – ANVISA), as the Food and Drug Administration (FDA) is in the United States.

In Brazil, the regulation that deals with the registration of a medicine is the RDC n° 24/2011, which has innovated by requesting complete information of the phytochemical and active supplies derived synthesis or semi-synthesis. Subsection II of Section V of Chapter III of this resolution stipulates that all necessary documents and studies be submitted by companies for the registration of a new drug, thereby ensuring the quality of the active principle, since pharmaceutical supplies produced by different manufacturers may have different impurities (solvent, enantiomers, polymorphs, degradation products, etc.) that may influence the stability of the medicament.

Medicines containing rutin and/or quercetin and/or hesperidin and/or diosmin and/or rutin and/or cumarin and those derived from synthesis or semi-synthesis of these compounds represent an exception to Art. 120 of the RDC n° 48/2009, in accordance with the provisions of single paragraph of the Art. 47 of the RDC n° 24/2011. Thus, even if it is necessary to carry out some procedure to alter the post-registration in these medicines, such as changing the synthetic route or manufacturing location, it must abide by the procedures laid out in the DRC No. 48/2009.

10. Hesperidin in Cosmetics and Skin Care

10.1. Use of Flavonoids in Cosmetics and Dermatology, Effects on the Skin

Phenolic acid, flavonoids, stilbenes, lignans, coumarins, and tannins are the most important polyphenols used in modern cosmetology and dermatology [61]. The antioxidant, anti-aging, anti-inflammatory, antimicrobial and anticancer properties of flavonoids are frequently deployed in skincare products. Their antiradical function, as well as the ability to inhibit the creation of enzymes and proteins, is one of their important effects. Flavonoids' high chemical reactivity can prevent injury caused by free radicals by the following mechanisms: direct scavenging of ROS, activation of antioxidant enzymes, chelating of metal compounds, reduction of α-tocopheryl radicals, inhibition of oxidases, reduction of oxidative stress caused by nitric oxide, increase in uric acid levels, and increase in antioxidant properties of low molecular antioxidant [58].

Flavonoids have an influence on skin microcirculation and can be used as ingredients in cream for vascular, oily and/or atopic skin [61]. Van der Waals forces, dipole-dipole and hydrogen bonding interaction, and the specific design of the compound, make them capable of reacting with other biomolecules, especially proteins. Dermal bioavailability and antioxidant activity of glycosides is lower compared to their aglycon forms, due to their low permeability [61].

Absorption of ultraviolet radiation is one of the chemical properties of flavonoids, based on the presence of the conjugated double bonds. These compounds may absorb UV radiation in the range of 240-285 nm and 300-550 nm, which is associated with the presence of the electron-donating substituents attached to the phenol rings, as well as intramolecular and intermolecular hydrogen bonds and steric effects. Polyphenol molecules can act as an exogenous chromophore or photosensitizer. After the absorption of quantum energy, they become excited and emit excess energy in the form of infrared radiation. Concentrations found in cosmetics cannot replace conventionally used synthetic UVA and UVB filters, but they can reduce erythema and skin burns caused by exposure to UVB radiation and IR rays of sunlight [61].

The anti-aging properties of flavonoids are associated with their effect on the modulation of protease activity, especially matrix metalloproteinase, which is dependent on the Zn^{2+} ion of enzymes involved in connective tissue remodeling.

A significant role is in their activity, either by sequestering metal ions or the effect they have on the expression of endogenous protein tissue inhibitors of metalloproteinase [61].

Flavonoids' skin-whitening activity is connected with their modulation of the tyrosinase activity through their chelating ability of Cu^{2+} ion in the formation of L-dopaquinone, which is part of the melanogenesis process. This ability is found for quercetin, cyaniding, and kaempferol, while the glycosidic forms have not shown this capability [61].

Dr Albert Szent-Gyorgi found that citrus peel flavonoids were effective in preventing capillary bleeding, and was the first to report on the biological activity of flavonoids on capillary fragility associated with scurvy [38]. Skin circulation may be improved by phenolic compounds through their protective effect on blood vessel walls and reduction of capillary permeability, and by facilitation of the free flow of blood through capillaries as a result of platelet aggregation. Inhibition of this process is associated with chelation of Ca^{2+} ions necessary for the proper operation of thromboplastin, catalyzing conversion of prothrombin into thrombin, which is in turn responsible for the conversion of fibrinogen to fibrin. Inhibition of hyaluronidase activity, the enzyme responsible for the enzymatic degradation of hyaluronic acid, which is a building block of the blood vessel wall and dermis, is one of the protective functions of the flavonoids [61].

Furthermore, they may also protect skin against exogenous, pathogenic bacteria, fungi, and viruses, inhibiting the activity of proteolytic enzymes by disturbing the synthesis of nucleic acids and increasing the permeability of the cell wall to exogenous substances. Additionally, phenolic compounds may have antimicrobial properties and assist in the preservation of cosmetics by protecting the physicochemical form of the product against secondary infection [61].

Apigenin, catechin, rutin, quercetin, luteolin, baicalin, hesperidin, and diosmin have the ability to prevent the aggregation of platelets in clot formation [61].

10.2. Hesperidin and Hesperetin in Cosmetics Uses

Hesperidin and its aglycon hesperetin are bioflavonoids with interesting bioactive properties that remain foci of intensive research for topical application. Sulfonated or phosphorylated hesperidin derivatives are extremely potent inhibitors of hyaluronidase, and acetylated hesperidin may cause inhibitory action of ascorbic acid on hyaluronidase. It has been found that hesperidin reduces superoxide in electron transfer as well as concerted proton transfer reaction *in vivo*. Hesperidin might be a topical photo-protective agent, by protecting phosphatidylcholine liposomes from UV-irradiation-induced peroxidation by scavenging oxygen free radicals generated by UV irradiation. It has also been shown to be able to penetrate through the *stratum corneum* [38].

In vitro study has shown that hesperidin and hesperetin penetrate the skin much faster in the presence of D-limonene and lecithin [79]. Transport of active ingredients through the *stratum corneum* occurs through the following steps: vehicle penetration to the *stratum corneum*, diffusion of active ingredients, and finally penetration below the *stratum corneum* in the living layers of epidermis. How quickly these processes occur depends on biological factors (age, skin condition, cardiovascular function and metabolism) and also physicochemical ones, such as the partition coefficient between the *stratum corneum* and the vehicle, and is related to lipophilicity as well as the size and spatial structure of the molecule, polarity and load. *In vitro* studies of *stratum corneum* have shown that individual flavonoids exhibit differences in penetration capacity, which depend in large part on the composition of the formulation in which the compound is present. It has been proven that the penetration rate of flavonoids (rutin, quercetin, and catechin) is affected by moisturizing substances (glycerol, glycols, polyglycols, ethoxylated methylglucoside, and urea) and also by cosmetic formulation (hydrogel, emulsion, microemulsion, and micellar system), or content of the mixture in which the active ingredients occur. Water/oil microemulsion formulation significantly enhances quercetin skin penetration 12 h after application [80]. Improvement of low drug permeability through human skin implies the use of penetration enhancers (including lipophilic solvents, surfactant, fatty acids and terpens), which reversibly modify the barrier resistance of the skin. Rutin, catechin, epicatechin, and quercetin have a limited penetration. Quercetin's poor permeability, for example, even in the presence of the enhancers, might be explained by its absolute insolubility in water [79]. Several authors suggest that the permeability of this compound can be increased by uploaded flavonoids in liposomes or other kinds of carrier systems. It can also be affected by the presence of promoters of permeability, such as glycerin and propylene glycol [61].

In the skin, hesperidin may significantly stimulate epidermal hyperplasia and improve epidermal permeability through epidermal proliferation, osteoblasts differentiation, and lipid secretion. Different skin layers and skin proteins can respond differently to hesperidin treatment. Involucrin, present in the *stratum spinosum*, does not respond to hesperidin treatment. Filaggrin and loricrin (located in *stratum granulosum* and *stratum spinosum)* expression is increased. These beneficial effects of improving epidermal permeability, barrier function and regulation of filaggrin can be useful in the treatment of certain skin disorders, such as cutaneous inflammation and atopic dermatitis [81]. *In vitro*, hesperidin can reduce the process of melanogenesis through the inhibition of tyrosinase activity in melanocytes, or through a reduction in melanocytes proliferation. Therefore, the possibility of affecting tyrosinase activity, and the practical application thereof in the field of anti-aging preparations designed to prevent the formation of senile lentigines (*lentigo senilis*) and solar lentigo (*lentigo solaris*), makes bioflavones very interesting for skin whitening [61]. Daily topical applications of hesperidin microemulsion have shown a significant skin whitening effect, reduction of trans-epidermal water loss and inhibition of irritation effect after exposure to UV-rays after 4 weeks [82]. In this study, Kim et al. showed that hesperidin had a depigmentation effect by blocking the melanophilin, the tripartite protein complex, which is responsible for transport of the melanosome into the melanocytes. This was proven by melanosome aggregation testing in cells. The hesperidin did not inhibit melanin production in melanoma cells, but reduced skin pigmentation in reconstruction of human epidermal skin [83]. Among other things, hesperidin may affect polyoxygenase, cyclooxygenase, hyaluronidase, collagenase, elastase, and tyrosinase, and can thus contribute to the reduction of modifications in the skin under the influence of enzymes involved in connective tissue remodeling [61]. Hyaluronidase plays a significant role in regulating the permeability of capillary walls and supporting tissues by causing the breakdown of hyaluronic acid and increasing the permeability of the tissue [38]. The ability to chelate multivalent metals such as Fe^{3+}, Fe^{2+}, Cu^{2+}, Zn^{2+} and Mn^{2+} (Figure 13) is as relevant to inhibition of enzymes, which contain metal ions in their reaction center or require metal cations as cofactors, as to inhibition of inflammatory processes and function of vascular vessels [61, 84].

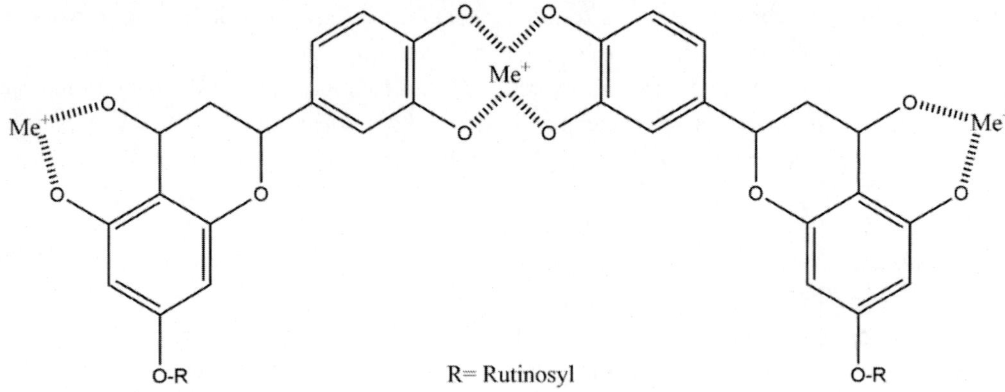

Figure 13. Hesperidin as a chelating agent of metal multivalent ions (Me^{+}). The creation of the complex involves OH groups in ortho- position that coordinate with divalent cations.

11. Other High Value-Added Products from Oranges

In addition to bioflavonoids, other high-value commodities such as pectin, second-generation (2G) bioethanol, and nanocellulose can be extracted from orange peel (Figure 14), adding even more value to the bagasse generated by the orange juice industry.

Pectin is a heterogeneous polysaccharide composed of galacturonic acid interspersed with methoxy galacturonic acid, whose side chain ramifications contain pentoses and hexoses, such as galactose, xylose, and/or arabinose. It is present in large amounts in the cell wall of oranges, thus improving cell wall chemical stability and physical strength [85]. Pectin is very abundant in the albedo, the white part of the orange, between the peel and the fruit pulp [86]. In its polymer form, pectin is mostly used in the food industry as a thickener for juices, lactic drinks, and jellies.

Second-Generation (2G) bioethanol produced from orange peel helps to supply increasing demand in the Brazilian and world market, especially with the rise of hybrid cars with engines adapted to both ethanol and gasoline [87].

The orange bagasse contains soluble sugars, such as glucose, fructose and sucrose, in addition to insoluble polysaccharides such as pectin, cellulose, and hemicellulose, which can be hydrolyzed using acids or enzymes to increase the amount of free sugars (monomers) released in the reaction medium.

Enzymatic hydrolysis is much a more advantageous method than acid hydrolysis because it demands less energy and not carry the risk of equipment corrosion.

However, this type of hydrolysis represents a high cost for companies, because commercial enzymes are very expensive, and for an industry that processes a large amount of bagasse this cost can derail the process [88].

Furthermore, the use of these biocatalysts presents a further series of disadvantages which impact the cost of biofuel, among which are: i) the need for large quantities of enzymes to decrease the time necessary to complete hydrolysis, ii) unproductive enzymes' adsorption on lignin molecules, which causes an enzyme performance loss, iii) enzymes inhibition in the presence of high concentrations of substrate or/and product (glucose), iv) the possibility of some enzyme activity loss due to its denaturation or degradation, or other factors linked to the substrate composition [89]. As a strategy to reduce the cost, the production of enzymes from different organisms has been enhanced [90, 91].

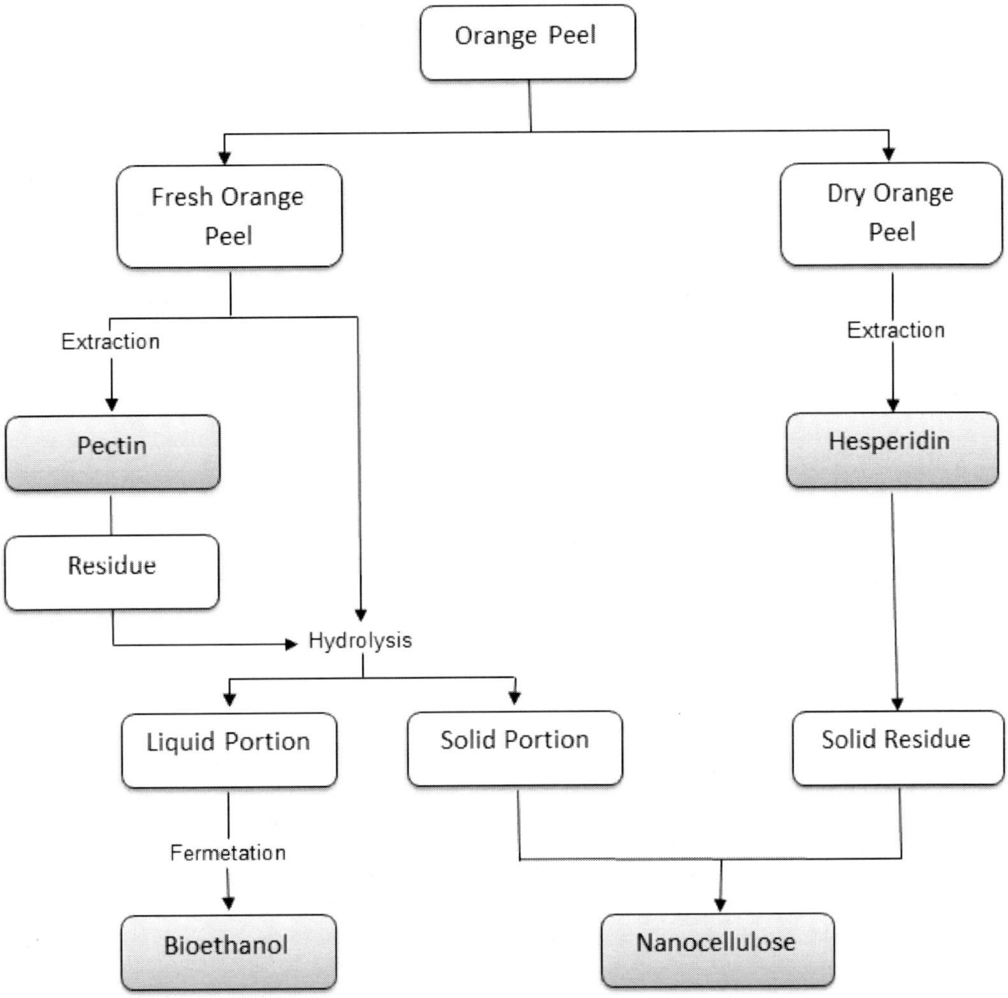

Figure 14. The most important and high-value commodities that can be obtained from orange peel after removing the orange juice: pectin, hesperidin, bioethanol and nanocellulose.

In addition, intensive research by companies and public research institutions has resulted in significant cost reduction to applying these enzymes in the process via different strategies, including: i) improvement of the specific activity levels, ii) reduction of cost production, and iii) increase of the enzymes' stability. Despite this progress, greater cost savings, either by improving the activity of enzymes produced by microorganisms such as fungi and bacteria, or through production by genetic engineering, are tasks, which are increasingly difficult [92]. Because of these obstacles, the process of second-generation ethanol production is still economically unfeasible, but the use and production of new enzymes can be the key to enable this production [88]. Within this context, our research group used the protein extract of the *Xanthomonas axonopodis* pv. *citri* strain 306, a gram-negative bacterial pathogen which causes citrus canker, as an alternative to commercial enzymes [90, 93]. The mechanism of bacterium action is to secrete enzymes such as cellulases and pectinases, which attack the orange for degrading its peel [94, 95, 96]. After hydrolysis, the liquid portion containing sugars is separated from the solid one and used in fermentation, an exothermic reaction in which the fermentable sugars are converted to bioethanol, as well as generating carbon dioxide and ATP. For this reaction to occur, it is necessary to use specific microorganisms, with certain species of yeasts and modified bacteria, being the most common in the use of *Saccharomyces cerevisiae* [97, 98].

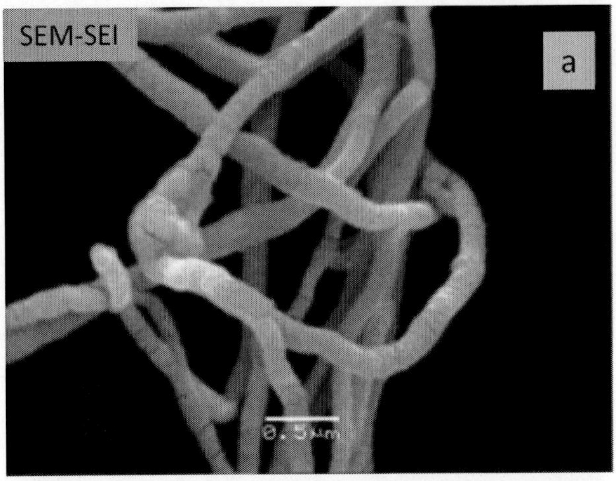

a

b

[a] Reproduced with permission from Reference [101].
[b] Reproduced with permission from Reference [82].

Figure 15. Scanning electron micrographs of nanocellulose obtained from citrus processing waste from oranges (CPWO) enzymatic hydrolysis: (a) nanocellulose obtained using commercial enzymes pulpzyme HA (Novozyme), celluclast 1.5 L. (Novozyme) and β-galactosidase from *Aspergillus oryzae* (Sigma). (b) nanocellulose obtained using *Xanthomonas axonopodis* pv. *citri* enzymes.

Nanocellulose can be obtained by defibrillation of the pulp fibers or partial hydrolysis of cellulose chains. For this purpose, mechanical, chemical, enzymatic, and ultrasound sonication methods, or a combination of these, can be used. In the plant cell wall, cellulose chains aggregate into the repeated crystalline and amorphous structure to form microfibrils, which also aggregate into larger macroscopic fibers [99]. Essentially, this hierarchical structure is deconstructed in order to generate cellulose nanofibers from plants. Initially, the amorphous regions of the cellulose chains are hydrolyzed due to their greater accessibility. When these amorphous regions are completely hydrolyzed, cellulose whiskers formed only by the crystalline regions are obtained. Whereas the designation "nanofibrils" should be used to designate long flexible nanoparticles consisting of alternating crystalline and amorphous strings, the term "whiskers" is used to designate elongated crystalline rod-like nanoparticles, obtained after total hydrolysis of amorphous regions [100].

The successful extraction of nanofibrils from orange peel has been performed by hydrolysis of this biomass "in natura" using commercial enzymes and cost-efficient enzymes from *Xanthomonas axonopodis* pv. citri *(Xac)*, respectively (Figure 15) [101, 102]. The nanocellulose isolated from the enzymatically treated solid residues has a long structure, with an average length of 458 ± 115 nm and a width of 10 ± 3 nm, generating an average aspect ratio of 47 ± 18 nm, which is considered high [102], and yields of 3% (g g-1 of dry orange peels) [101].

CONCLUSION

As this paper has demonstrated, orange peel has very interesting ingredients with fascinating properties, and should be used more in a human diet. Orange peel may also be used as an excellent source for isolating hesperidin and hesperetin, not currently used in great quantities but which may have very useful application in avoiding, preventing, and/or helping to cure various human health problems. In addition, as well as bioflavonoids, other high-value commodities, such as pectin, second-generation (2G) bioethanol, and nanocellulose can be extracted from orange peel, and may contribute to a greener and more sustainable way of dealing with agricultural waste from the orange juice industry worldwide.

Authors' contributions and conflict of interest: This chapter has been designed and structured by LT. All authors wrote the chapter and participated in revising the draft. Also all authors have contributed with graphics, materials and/or analysis tools. The authors of this chapter declare no conflicts of interest.

ACKNOWLEDGMENTS

We would like to kindly acknowledge CAPES, OPCW, FAPESP and CNPq for financial support and fellowships.

REFERENCES

[1] Di Giacomo, A., Dugo, G. *Citrus: The Genus Citrus*. US/Canada: Taylor and Francis; 2002.

[2] Rekha, S. S., Bhaskar, M. In vitro screening and identification of antioxidant activities of orange (Citrus sinensis) peel extract in different solvents. *International Journal of Pharma and Bio Sciences*, 2013, 4, 405-412.

[3] Orwa, C., Mutua, A., Kindt, R., Jamnadass, R., Anthony, S. *Agroforestree Database: a tree reference and selection guide version 4.0*. Kenya: World Agroforestry Centre, 2009.

[4] Ikan, R. *Natural Products: A Laboratory Guide*. 2nd Edition. US: Academic Press; 1991.

[5] Ladaniya, M. S. *Citrus Fruit: Biology, Technology and Evaluation*; 1st Edition. Elsevier Inc. 2007.

[6] Srivastava, A. K. *Advances in Citrus Nutrition*; 1st Edition. Springer Science + Business Media B.V.; 2012.

[7] Seitz, C. T., Wingard Jr., R. E. An improved conversion of hesperidin into hesperetin including purity determination by gradient-elution, high-pressure liquid chromatography; *Journal of Agricultural and Food Chemistry*; 26, 278-280; 1968.

[8] Raw, L. Ramachandra, Sastry, P. G. Glycosides of citrus fruits. I. Oranges *Indian Journal of Applied Chemistry*, 1962, 25, 86-9.

[9] Yu, L., Huang, H., Yu, L. L., Wang, T. T. Y. Utility of hesperidinase for food function research: enzymatic digestion of botanical extracts alters cellular antioxidant capacities and anti-inflammatory properties. *Journal of Agricultural and Food Chemistry*, 2014, 62, 8640-8647.

[10] M'hiri, N., Ioannou, I., Mihoubi Boudhriou, N., Ghoul, M. Effect of different operating conditions on the extraction of phenolic compounds in orange peel. *Food and bioproducts processing*; 96, 161-170; 2015.

[11] Rezzadori, K., Benedetti, S., Amante, E. R. Proposals for the residues recovery: Orange waste as raw material for new products Food and bioproducts processing. *Food Bioprod. Process*. 2012, 90, 606-614.

[12] Manach, C., Morand, C., Gil-Isquierdo, A., Bouteloup-Demange, C., Rémésy, C. Bioavailability in humans of the flavones hesperitin and narirutin after the ingestion of two doses of orange juice. *European Journal of Clinincal Nutrition*, 2003, 57, 235-242.

[13] Zubrick, J. W. *The Organic Chem Lab Survival Manual*, 4th Edition, New York: Wiley and Sons, Inc., 1997.

[14] Sharma, P., Pandey, P., Gupta, R., Roshan, S., Garg, A., Shulka, A., Pasi, A. Isolation and characterization of hesperidin from orange peel. *Indo American Journal of Pharmaceutical Research*, 2013, ISSN NO: 2231-6876.

[15] Hendrickstone, R., Kesterstone, J. W. Purification of crude hesperidin. *Florida state horticultural society*. 1955, 3, 121-124.

[16] Londoño-Londoño, J., Rodrigues de Lima, V., Lara, O., Gil, L., Crecsynski Pasa, T. B., Arango, G. J., Ramirez Pineda, J. Clean recovery of antioxidant flavonoids from citrus peel: Optimizing an aqueous ultrasound-assisted extraction method. *Food Chem*. 2010, 119, 81-87.

[17] Wang, A. Y., Zhou, M. Y., Lin, W. C., Antioxidative and anti-inflammatory properties of Citrus sulcata extracts. *Food Chem.*, 2011, 124, 958-963.

[18] Inouea, T., Tsubakia, S., Ogawab, K., Onishic, K., Azuma, J. Isolation of hesperidin from peels of thinned Citrus unshiu fruits by microwave-assisted extraction. *Food Chem.*, 2010, 123, 542-547.

[19] Li, B. B., Smith, B., Hossain, M. Extraction of phenolics from citrus peels. I. Solvent extraction method. *Sep. Purif. Technol.* 2006, 48, 182-188.

[20] M'hiria, N., Ioannou, I., Ghoul, M., Mihoubi Boudhrioua, N. Extraction Methods of Citrus Peel Phenolic Compounds. *Food Reviews International*, 2014, 30, 265-290.

[21] Jeong, S. M., Kim, S. Y., Kim, D. R., Jo, S. C., Nam, K. C., Ahn, D. U., Lee, S. C. Effect of heat treatment on the antioxidant activity of extracts from citrus peels. *J. Agric. Food Chem.* 2004, 52(11), 3389-3393.

[22] Cheigh, C. I., Chung, E. Y., Chung, M. S. Enhanced extraction of flavanones hesperidin and narirutin from Citrus unshiu peel using subcritical water. *J. Food Eng.* 2012, 110, 472-477.

[23] Khan, M. K., Abert-Vian, M., Fabiano-Tixier, A. S., Dangles, O., Chemat, F. Ultrasound-assisted extraction of polyphenols (flavanone glycosides) from orange (Citrus sinensis L.) peel. *Food Chem.* 2010, 119, 851-858.

[24] Ma, Y. Q., Chen, J. C., Liu, D. H., Ye, X. Q. Effect of ultrasonic treatment on the total phenolic and antioxidant activity of extracts from citrus peel. *J. Food Sci.* 2008, 73(8), 115-120.

[25] Lewandowska, U., Szewczyk, K., Hrabec, E., Janecka, A., Gorlach, S. Overview of metabolism and bioavailability enhancement of polyphenols. *Journal of Agricultural and Food Chemistry*, 2013, 61, 12183-12199.

[26] Manach, C., Williamson, G., Morand, C., Scalbert, A., Rémésy, C. Bioavailability and bioefficacy of polyphenols in humans. I. Review of 97 bioavailability studies. *American Journal of Clinical Nutrition*, 2005, 81, 230S-242S.

[27] Grohmann, K., Manthey, J. A., Cameron, R. G. Acid-catalyzed hydrolysis of hesperidin at elevated temperatures. *Carbohydrate Research*, 2000, 328, 141-146.

[28] Lahmer, N., Belboukhari, N., Cheriti, A., Sekkoum, K. Hesperidin and hesperitin preparation and purification from *Citrus sinensis* peels. *Der Pharma Chemica*, 2015, 7, 1-4.

[29] Shin, K. C., Nam, H. K., Oh, D. K. Hydrolysis of flavanone glycosides by β-glucosidase from *Pyrococcus furiosus* and its application to the production of flavanone aglycones from citrus extracts. *Journal of Agricultural and Food Chemistry*, 2013, 62, 11532-11540.

[30] Battestin, V., Macedo, G. A., De Freitas, V. P. A. Hydrolysis of epigallocatechin gallate using a tannase from *Paecilomyces variotii. Food Chemistry*, 2008, 108, 228-233.

[31] Macedo, J. A., Battestin, V., Ribeiro, M. L., Macedo, G. A. Increasing the antioxidant power of tea extracts by biotransformation of polyphenols. *Food Chemistry*, 2011, 126, 491-497.

[32] Dewick, M. P. *Medicinal Natural Products, a Biosynthetic Approach.* 2nd Edition. University of Nottingham, UK: John Wiley and Sons; 2002.

[33] Ferrer, J. L., Austin, M. B., Stewart, C. Jr; Noel, J. P., Structure and function of enzymes involved in the biosynthesis of phenylpropanoids. *Plant Physiol. Biochem.*, 2008, 46(3), 356-70.

[34] Saänches-Morreno, C., Plaza, L., Elez-Martiänez, P., De Ancos, B., Martiän-Belloso, O., Cano, M. P. Impact of high pressure and pulsed electric fields onbioactive compounds and antioxidant activity of orange juicein comparison with traditional thermal processing. *J. Agric. Food Chem.* 2005, 53, 4403-4409.

[35] Kuntić, V., Brborić, J., Holclajtner-Antunović, I., Usković-Marković, S. Evaluating the bioactive effects of flavonoid hesperidin – a new literature data survey. *Vojnosanitetski Pregled*, 2014, 71, 60-65.

[36] Harman, D. Free radical theory of aging: the "free radical" diseases; *Age*. 1984, *7*, 111-131.

[37] Saija, A., Scalese, M., Lanza, M., Marzullo, D., Bonina, F., Castelli, F. Flavonoids as antioxidant agents: importance of their interaction with biomembranes. *Free Radical Biology and Medicine*, 1995, 19, 481-486.

[38] Garg, A., Garg, S., Zaneveld, L. J. D., Singla, A. K. Chemistry and pharmacology of the citrus bioflavonoid Hsd, *Phytotherapy Research*, 2001, 15, 655-669.

[39] Heim, K. E., Tagliaferro, A. R., Bobilya, D. J. Flavonoid antioxidants: chemistry, metabolism and structure-activity relationships. *Journal of Nutritional Biochemistry*, 2002, 13, 572-584.

[40] Parhiz, H., Ali, R., Soltani, F., Rezaee, R., Iranshahi, M. Antioxidant and anti-inflammatory properties of the citrus flavonoids hesperidin and hesperetin: an updated review of their molecular mechanisms and experimental models. *Phytotherapy Research*, 2015, 29, 323-331.

[41] Jovanovic, S. V., Steenken, S., Tosic, M., Marjanovic, B., Simic, M. G. Flavonoids as antioxidants; *Journal of the American Chemical Society*, 1994, 116, 4846-4851.

[42] Chen, M. C., Ye, Y. Y., Ji, G., Liu, J. W. Hesperidin upregulates heme oxygenase-1 to attenuate hydrogen peroxide-induced cell damage in hepatic L02 cells. *Journal of Agricultural and Food Chemistry*, 2010, 58, 3330-3335.

[43] Elavarasan, J., Velusamy, P., Ganesan, T., Ramakrishnan, S. K., Rajasekaran, D., Periandavan, K. Hesperidin-mediated expression of Nrf2 and upregulation of antioxidant status in senescent rat heart. *Journal of Pharmacology and Pharmacotherapeutics*, 2012, 64, 1472-1482.

[44] Martinez, M. C., Fernandez, S. P., Loscalzo, L. M., Wasowski, C., Paladini, A. C., Marder, M., Medina, J. H., Viola, H. Hesperidin, a flavonoid glycoside with sedative effect, decreases brain pERK1/2 levels in mice. *Pharmacoogy Biochemistry and Behaviour*, 2009, 92, 291-296.

[45] Wilmsen, P. K., Spada, D. S., Salvador, M. Antioxidant activity of the flavonoid hesperidin in chemical and biological systems. *Journal of Agricultural and Food Chemistry*, 2005, 53, 4757-4761.

[46] Skoog, D. A., West, D. M., Holler, J. F. *Fundamentals of Analytical Chemistry*. 9th edition. Belmont, US: Brooks/Cole; 2004.

[47] Yin, Y., Li, Z., Zhong, Z., Gates, B., Xia, Y., Venkateswaran, S. Synthesis and characterization of stable aqueous dispersions of silver nanoparticles through the Tollens process. *Journal of Materials Chemistry*, 2002, 12, 522-527.

[48] Duda-Chodak, A. The inhibitory effect of polyphenols on human gut microbiota. *Journal of Physiology and Pharmacology*, 2012, 63, 497-503.

[49] Assini, J. M., Mulvihill, E. E., Huff, M. W. Citrus flavonoids and lipid metabolism. *Current Opinion in Lipidology*, 2013, 24, 34-40.

[50] Gao, S., Jiang, W., Yin, T., Hu, M. Highly variable contents of phenolics in St. John's Wort products affect their transport in the human intestinal caco-2 cell model: Pharmaceutical and biopharmaceutical rationale for product standardization. *Journal of Agricultural and Food Chemistry*, 2010, 58, 6650-6659.

[51] Tanaka, T., Makita, H., Kawabata, K., Mori, H., Kakumoto, M., Satoh, K., Hara, A., Sumida, T., Tanaka, T., Ogawa, H. Chemoprevention of azoxymethane-induced rat colon carcinogenesis by the naturally occurring flavonoids, diosmin and hesperidin. *Carcinogenesis*, 1997, 18, 957-965.

[52] Tanaka, T., Makita, H., Ohnishi, K., Mori, H., Satoh, K., Hara, A., Sumida, T., Fukutani, K., Tanaka, T., Ogawa, H. Chemoprevention of 4-nitroquinoline1-oxide induced oral carcinogenesis in rats by flavonoids diosmin and hesperidin, each alone and in combination. *Cancer Research*, 1997, 57, 246-252.

[53] So, F. V., Guthrie, N., Chambers, A. F., Carroll, K. K. Inhibition of proliferation of estrogen receptor-positive MCF-7 human breast cancer cells by flavonoids in the presence and absence of excess estrogen. *Cancer Letters*, 1997, 112, 127-133.

[54] Lentini, A., Forni, C., Provenzano, B., Beninati, S. Enhancement of transglutaminase activity and polyamine depletion in B16-F10 melanoma cells by flavonoids naringenin and hesperetin correlate to reduction of the *in vivo* metastatic potential. *Amino Acids*, 2007, 32, 95-100.

[55] Sambantham, S., Radha, M., Paramasivam, A., Anandan, B., Malathi, R., Chandra, S. R., Jayaraman, G. Molecular mechanism underlying hesperetin-induced apoptosis by *in silico* analysis and in prostate cancer PC-3 cells. *Asian Pacific Journal of Cancer Prevention*, 2013, 14, 4347-4352.

[56] Sivagami, G., Vinothkumar, R., Preethy, C. P., Riyasdeen, A., Akbarsha, M. A., Menon, V. P., Nalini, N. Role of hesperetin (a natural flavonoid) and its analogue on apoptosis in HT-29 human colon adenocarcinoma cell line – a comparative study. *Food and Chemical Toxicology*, 2012, 50, 660-671.

[57] Zhang, J., Song, J., Wu, D., Wang, J., Dong, W. Hesperetin induces the apoptosis of hepatocellular carcinoma cells via mitochondrial pathway mediated by the increased intracellular reactive oxygen species, ATP and calcium. *Medical Oncology*, 2015, 32-101.

[58] Roohbakhsh, A., Parhiz, H., Soltani, F., Rezaee, R., Iranshahi, M. Molecular mechanisms behind the biological effects of hesperidin and hesperetin for prevention of cancer and cardiovascular diseases. *Life Sciences*, Elsevier Inc. 2014, 124, 67-74.

[59] Chanet, A., Milenkovic, D., Manach, C., Mazur, A., Morand, C.; Citrus Flavanones: What Is Their Role in Cardiovascular Protection?; *J. Agric. Food Chem.* 2012, 60, 8809-8822.

[60] Fraga, C. G. *Plant Phenolics and Human Health, Biochemistry, Nutrition, and Pharmacology.* New Jersey: Wiley and Sons, Inc.; 2010.

[61] Ratz-Lyko, A., Arct, J., Majewski, S., Pytkowska, K. Influence of polyphenols on the physiological processes in the skin. *Phytotherapy Research*, 2015, 29, 509-517.

[62] Guardia, T., Rotelli, A. E., Juarez, A. O., Pelzer, L. E. Anti-inflammatory properties of plant flavonoids. Effects of rutin, quercetin and hesperidin on adjuvant arthritis in rat. *Il Farmaco*, 2001, 56, 683-687.

[63] Rotelli, A. E., Juarez, A. O., Guardia, T., La Rocha, N. E., Pelzer, L. E. Comparative study of flavonoids in experimental models of inflammation. *Pharmacology Research*, 2003, 48, 601-606.

[64] Pan, M. H., Lai, C. S., Ho, C. T. Anti-inflammatory activity of natural dietary flavonoids. *Food and Function*, 2010, 1, 15-31.

[65] Jain, M., Parmar, H. S. Evaluation of antioxidative and anti-inflammatory potential of hesperidin and naringin on the rat air pouch model of inflammation. *Inflammation Research*, 2011, 60, 483-491.

[66] Vallejo, F., Larrosa, M., Escudero, E., Zafrilla, M. P., Cerda, B. A., Boza, J., Garcia-Conesa, M. T., Espin, J. C., Toma, F. A. Concentration and solubility of flavanones in orange beverages affect their bioavailability in humans. *Journal of Agricultural and Food Chemistry*, 2010, 58, 6516-6524.

[67] Mullen, W., Archeveque, M. A., Edwards, C., Matsumoto, H., Crozier, A. Bioavailability and metabolism of orange juice flavanones in humans: impact of a full-fat yogurt. *Journal of Agricultural and Food Chemistry*, 2008, 56, 11157-11164.

[68] Mullen, W., Edwards, C. A., Serafini, M., Crozier, A. Bioavailability of pelargonidin-3-glucoside and its metabolites in humans following the ingestion of strawberries with and without cream. *Journal of Agricultural and Food Chemistry*, 2008, 56, 713-719.

[69] Yamamoto, M., Jokura, H., Hashizume, K., Ominami, H., Shibuya, Y., Suzuki, A., Hase, T., Shimotoyodome, A. Hesperidin metabolite hesperitin-7-O-glucoronide, but not hesperitin-3'-O-glucoronide, exerts hypertensive, vasodilatory, and anti-inflammatory activities. *Food and Function*, 2013, 4, 13460-1351.

[70] Erlund, I., Meririnne, E., Alfthan, G., Aro, A. Plasma Kinetics and Urinary Excretion of the Flavanones Naringenin and Hesperetin in Humans after Ingestion of Orange Juice and Grapefruit Juice. *Journal of Nutrition*, 2001, 131, 235-241.

[71] Brand, W., Boersma, M. G., Bik, H., Hoek-van den Hil, E. F., Vervoort, J., Barron, D., Meinl, W., Glatt, H., Williamson, G., Van Bladeren, P. J., Rietjens, I. M. C. M. Phase II metabolism of hesperetin by individual UDP Glucuronosyltransferases and Sulfotransferases and rat and human tissue samples. *Drug Metabolism and Disposition*, 2010, 38, 617-625.

[72] Nielsen, I. L., Chee, W. S., Poulsen, L., Offord-Cavin, E., Rasmussen, S. E., Frederiksen, H., Ensten, M., Barron, D., Horcaiada, M. N., Williamson, G. Bioavailability is improved by enzymatic modification of the citrus flavonoid hesperidin in humans: a randomized, double-blind, crossover trial. *Journal of Nutrition*, 2006, 136, 404-408.

[73] Han, B., Yang, B., Yang, X., Zhao, Y., Liao, X., Gao, C., Wang, F., Jiang, R. Host-guest inclusion system of norathyriol with β-cyclodextrin and its derivatives: preparation, characterization, and anticancer activity. *Journal of Bioscience and Bioengineering*, 2014, 117, 775-779.

[74] Corciova, A., Lazar, D., Lazar, M. Spectral (UV-Vis, IR, NMR) and thermal (Tg, DTG, ATD) characterization of some synthesis brom-isoflavanone and the inclusion compounds with beta-cyclodextrin. *Revista de Chimie*, 2010, 61, 377-379.

[75] Petry, M., Borghetti, G. S., Bassani, V. L. Influência de ciclodextrinas e polímero hidrofílico sobre a hidrossulubilidade de diferentes formas polimórficas de quercetina. *Latin American Journal of Pharmacy*, 2007, 26, 831-836.

[76] Zhang, L., Zhang, Z., Li, N., Wang, N., Wang, Y., Tang, S., Xu, L., Ren, Y. Synthesis and evaluation of a novel β-cyclodextrin derivative for oral insulin delivery and absorption. *International Journal of Biological Macromolecules*, 2013, 61, 494-500.

[77] Shulman, M., Cohen, M., Soto-Gutierrez, A., Yagi, H., Wang, H., Goldwasser, J., Lee-Parsons, C. W., Benny-Ratsaby, O., Yarmush, M. L., Nahmias, Y. Enhancement of Naringenin Bioavailability by Complexation with Hydroxypropoyl-b-Cyclodextrin. *Plos one*, 2011, 6, 1-8.

[78] Miyake, K., Arima, H., Hirayama, F., Yamamoto, M., Horikawa, T., Sumiyoshi, H., Noda, S., Uekama, K. Improvement of solubility and oral bioavailability of rutin by complexation with 2-hydroxypropyl-b-cyclodextrin. *Pharmaceutical Development and Technology*, 2000, 5, 399-407.

[79] Saija, A., Tomaino, A., Trombetta, D., Giacchi, M., Pasquale, A. de; Bonina, F.; Influence of different penetration enhancers on in vitro skin permeation and in vivo photoprotective effect of flavonoids. *Int. J. Pharm.* 1998. 175, 85-94.

[80] Vicentini, F., Simi, T., Del Ciampo, J., Wolga, N., Pitol, D., Iyomasa, M., Bentley, V., Fonseca, M. Quercetin in w/o microemulsion: in vitro and in vivo skin penetration and efficacy against UVB-induced skin damages evaluated in vivo. *European Journal of Pharmaceutics and Biopharmaceutics*, 2008, 69, 948-957.

[81] Hou, M., Man, M., Man, W., Zhu, W., Hupe, M., Park, K., Crumine, D., Elias, P. M., Man, M. Topical hesperidin improves permeability barrier function and epidermal differentiation in normal murine skin. *Experimental Dermatology*, 2012, 21, 337-340.

[82] Yi-Hung, T., Ko-Feng, L., Yaw-Bin, H., Chi-Te, H., Pao-Chu, W. In vitro permeation and in vivo whitening effect of topical hesperetin microemulsion delivery system. *International Journal of Pharmaceutics*, 2010, 388, 257-262.

[83] Kim, B., Lee, J., Lee, H., Nam, K., Park, J., Lee, S., Kim, J., Lee, J., Hwang, J. Hesperidin suppresses melanosome transport by blocking the interaction of Rab27A-Melanophilin. *Biomolecules and Therapeutics*, 2013, 21, 343-348.

[84] An, B., Kwak, J., Park, J., Lee, J., Park, T., Son, J., Jo, C., Byun, M. Inhibition of enzyme activities and the antiwrinkle effect of polyphenol isolated from the persimmon leaf (Diospyros kaki folium) on human skin. *Dermatologic Surgery*, 2005, 31, 848-854.

[85] Willats, W. G. T., Knox, P. J., Mikkelsen, J. D. Pectin: new insights into an old polymer are starting to gel. *Trends in Food Science and Technology*, 2006, 17, 97-104.

[86] Mussatto, S. I., Teixeira, J. A. Lignocellulose as raw material in fermentation processes. In: Méndez-Vilas, A., editor. *Current research, technology and education topics in applied microbiology and microbial biotechnology*. Badajoz, Spain: Formtex; 2010, 897-907.

[87] Leite, R. C. C. Bioetanol combustível: uma oportunidade para o Brasil [online]. 2009 [cited 2014 08 19]. Available from http://www.cogen. com.br/paper/2010/Livro_ Bioetanol_Cana_Acucar_2009.pdf.

[88] Rodrigues, A. C., Leitão, A. F., Moreira, S., Felby, C., Gama, M. Recycling of cellulases in lignocellulosic hydrolysates using alkaline elution. *Bioresource Technology*, 2012, 110, 526-533.

[89] Jorgensen, H., Kristensen, J. B., Felby, C. Enzymatic conversion of lignocellulose into fermentable sugars: challenges and opportunities. *Biofuels, Bioproducts and Biorefining*, 2007, 1, 119-134.

[90] Awan, A. T., Tsukanomoto, J., Tasic, L. Orange waste as a biomass for 2G-ethanol production using low cost enzymes and co-culture fermentation. *RSC Advances*, 2013, 3, 25071-25078.

[91] Himmel, M. E., Ding, S. Y., Johnson, D. K., Adney, W. S., Nimlos, M. R., Brady, J. W., Foust, T. D. Biomass Recalcitrance: Engineering Plants and Enzymes for Biofuels Production. *Science*, 2007, 315, 804-807.

[92] Zhang, Y. H. P., Himmel, M. E., Mielenz, J. R. Outlook for cellulose improvement: Screening and selection strategies. *Biotechnology Advances*, 2006, 24, 452-481.

[93] INPI, Tasic, L., Tsukamoto, J., Awan, A. T., Durán, N. Processo de obtenção de bioetanol, esperidina e nanocelulose a partir de bagaço de laranja. BR1020130325856, 2013.

[94] asic, L., Borin, P. F. L., Khater, L., Ramos, C. H. I. Cloning and characterization of three hypothetical secretion chaperone proteins from *Xanthomonas axonipodis* pv. *Citri. Protein Expression and Purification*, 2007, 53, 363-369.

[95] Khater, L., Alegria, M. C., Borin, P. F. L., Santos, T. M., Docena, C., Tasic, L., Farah, C. S., Ramos, C. H. I. Identification of the flagellar chaperone FlgN in the phytopathogen Xanthomonas axonopodis pathovar citri by its interaction with hook-associated FlgK. *Archives of Microbiology*, 2007, 188, 243-250.

[96] Fattori, J., Prando, A., Assis, L. H. P., Aparicio, R., Tasic, L. Structural insights on two hypothetical secretion chaperones from Xanthomonas axonopodis pv. *citri. Protein Journal*, 2011, 30, 324-333.

[97] Madhavan, A., Tamalampudi, S., Srivastava, A., Fukuda, H., Bisaria, V. S., Kondo, A. Alcoholic fermentation of xylose and mixed sugars using recombinant Saccharomyces cerevisiae engineered for xylose utilization. *Applied Microbiology and Biotechnology*, 2009, 82, 1037-1047.

[98] Widmer, W., Zhou, W., Grohmann, K. Pretreatment effects on orange processing waste for making ethanol by simultaneous saccharification and fermentation. *Bioresource Technology*, 2010, 101, 5242-5249.

[99] Kalia, S., Kaith, B. S., Kaur, I. *Cellulose fibers: bio- and nano-polymer composites: green chemistry and technology*. London/New York: Springer-Verlag Berlin Heidelberg; 2011.

[100] Eichhorn, S. J., Dufresne, A., Aranguren, M., Marcovich, N. E., Capadona, J. R., Rowan, S. J., Weder, C., Thielemans, W., Roman, M., Renneckar, S., Gindl, W., Veigel, S., Keckes, J., Yano, H., Abe, K., Nogi, M., Nakagaito, A. N., Mangalam, A., Simonsen, J., Benight, A. S., Bismarck, A., Berglund, L. A., Peijs, T. Current international research into cellulose nanofibres and nanocomposites. *Journal of Materials Science*, 2010, 45, 1-33.

[101] Tsukamoto, J., Durán, N., Tasic, L. Nanocellulose and Bioethanol Production from Orange Waste using Isolated Microorganisms. *Journal of the Brazilian Chemical Society*, 2013, 24, 1537-1543.

[102] Mariño, M., Da Silva, L. L., Durán, N., Tasic, L. Enhanced Materials from Nature: Nanocellulose from Citrus Waste. *Molecules*, 2015, 20, 5908-5923.

In: Citrus Fruits
Editor: Daphne Simmons
ISBN: 978-1-63484-078-1

Chapter 3

PHYSIOLOGICAL PROPERTIES OF NARIRUTIN AND HESPERIDIN ISOLATED FROM *CITRUS UNSHIU*

Ho-Young Park and Inwook Choi*
Division of Functional Food Research, Korea Food Research Institute,
Gyeonggi, Republic of Korea

ABSTRACT

Citrus unshiu is one of the most important varieties of citrus grown in Northeast Asia. Its peel is known as 'Chinpi,' a non-toxic edible ethnopharmaceutical herb in China and Korea, and has been clinically used as a traditional medicine to treat common cold, dyspepsia, cough and phlegm. Modern therapeutic studies have proven that citrus flavonoids have anti-oxidative, anti-inflammatory and anti-allergic activities. In this chapter, an efficient way to isolate citrus flavonoids, narirutin and hesperidin, from *Citrus unshiu* was introduced. Physiological properties such as anti-inflammatory activities and anti-alcoholic liver disease were also reviewed with suggestions on improving their bioavailability in a body through enzymatic modifications.

1. INTRODUCTION

Citrus fruits have been cultivated widely around the world and include orange, lemons, limes, grapefruits and tangerines. Current annual worldwide citrus production is estimated over 115 million tons, with more than half as oranges [1]. It has been well established that citrus and citrus products are rich sources of vitamins, dietary fiber, and minerals, which are considered to play important roles in preventing diseases such as cardiovascular disease and cancer. Therefore, consumption of citrus fruits has grown as the public has more concern about their health than ever before.

* Corresponding author at: Division of Functional Food Research, Korea Food Research Institute, 516 Baekhyun-dong, Bundang-gu, Gyeonggi 463-746, Republic of Korea. Tel./Fax: +82-31-780-9097/+82-31-709-9876. E-mail address: choiw@kfri.re.kr (I. Choi).

Citrus fruits typically contain a little more than 35 mg of vitamin C per 100 g of flesh. Vitamin C is a potent antioxidant and prevents damages from free radicals. It also plays an important role in synthesizing collagen. Therefore, we can expect that consumption of vitamin C-rich foods may help decrease the risk of many diseases such as osteoarthritis, cancer, heart disease, and hypertension.

Citrus fruits are also good sources of soluble fiber that swell with water in the digestive tract to form a viscous mass. A diet with high amounts of dietary fiber can help prevent diabetes and high cholesterol and may lessen risk of stroke, obesity, hypertension, ulcers, and hemorrhoids.

Citrus fruits also contain abundant minerals. Main minerals (K, Ca, Na and Mg) and trace elements (Cu, Fe, Mn and Zn) are typically higher in peels than in pulps [2]. Potassium was the most abundant element presented, followed by Ca and Mg. Together, these minerals contribute to the synthesis of red blood cells and bone, and support the immune system. Thus, the ratio of K and Na in citrus plays an important role in maintaining the electrolyte balance of human cells [3].

Other than these biofunctional compounds, citrus fruits are rich sources of phytochemicals that have a wide range of physiological effects and help to prevent chronic diseases. Several classes of phytochemicals, including carotenoids, limonoids and flavonoids in citrus fruits have been reported.

2. CITRUS FLAVONOIDS

Flavonoids are one of the most common polyphenolic molecules, and they are abundantly present in foods such as fruits, vegetables, and plant-derived juice including tea and wine [4]. They exist in plants and vegetables as secondary metabolites and many studies have proved a relationship between the consumption of flavonoid-rich diets and prevention of human disease including cancer [5, 6], osteoporosis [7], cardiovascular disease [8], type-2 diabetes [9], and neurodegenerative disorders [10]. There are several classes of flavonoids: flavones (luteolin and apitenin), flavonols (quercetin and kaempferol), flavanones (hesperidin and narirutin), flavanonols (taxifolin and dihydrokaempferol), and flavans (catechin and gallocatechin), which characterize these molecules based on their carbon structure. These flavonoids are more concentrated in peels of citrus than flesh. Mouly et al. reported flavanone glycoside composition of citrus such as lemon, lime, grapefruit, and sweet orange [11]. Hesperidin, naringin and narirutin are found as main citrus flavonoids (Figure 1).

Hesperidin, hesperetin-7-*O*-rutinoside, was isolated for the first time from albedo of granges by Lebreton in 1827 and has since been found in lemons and other citrus fruits. Preclinical studies and clinical trials of hesperidin demonstrated therapeutic effects in various diseases, such as cardiovascular disease, neurological disorders, and others, due to its anti-inflammatory, antioxidant, and lipid-lowering properties [12].

Narirutin, naringein-7-*O*-rutinoside, is found abundantly in flesh and peel of citrus fruits like tangerine and grape fruits and reported to have anti-oxidative, anti-inflammatory, anti-allergic, anti-atopic dermatitis, and oviposition-stimulatory activities [13].

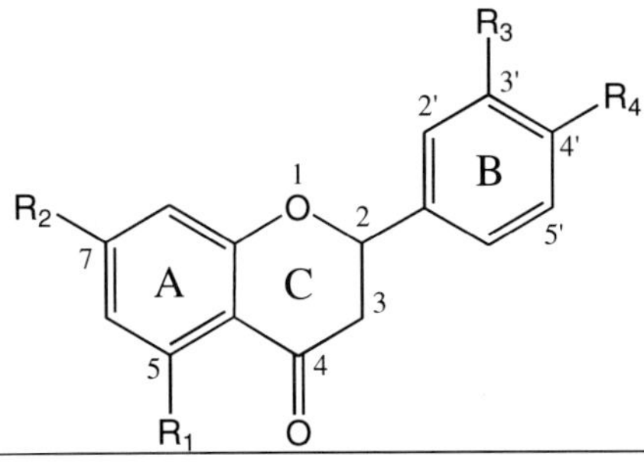

	R_1	R_2	R_3	R_4
Hesperidin	OH	[a](R-G)-*O*-	OH	OCH_3
Naringin	OH	[b](R-G)-*O*-	H	OH
Narirutin	OH	[a](R-G)-*O*-	H	OH

[a]Rutinoside (6-*O*-α-L-rhamnosyl-D-glucose).
[b]Neohesperidoside (2-*O*-α-L-rhamnosyl-D-glucose).

Figure 1. 2-D structures of narirutin and hesperidin.

Naringin, naringein-7-*O*-neohesperidoside, is another main flavonoid in citrus fruits, especially grapefruit (53 mg/100g) [14], which is responsible for the fruit's bitter taste. Naringin reduces bioavailability of many pharmacological agents taken at the same time. Naringin is metabolized into its aglycone, naringenin, in the human body. In spite of several bioactive activities, only 4% and 8% of ingested naringenin and naringin will get absorbed into the gastrointestinal tract [15, 16], respectively. These naringin and naringenin are antioxidants [17] and anti-inflammatory agents [18], otherwise naringenin is more potent compared with naringin because the sugar moiety (rhamnoglucoside) of naringin causes steric hindrance of the scavenging group [19].

3. Narirutin and Hesperidin from Citrus Peels

About 20% of citrus fruits are processed for citrus juice or other processed foods, and roughly 20% of their total weight is recovered as citrus byproducts. Citrus peels are the main components of citrus byproducts and rich sources of bio-functional flavonoids. Therefore, they could be good resources of bio-functional flavonoids [20, 21].

C*itrus unshiu* is widely cultivated in Northeast Asia and its peel, which is known as *Chinpi*, has been traditionally considered as a non-toxic edible ethnopharmaceutical herb and clinically used as a medicine for common cold, dyspepsia, cough and phlegm [22]. Narirutin, along with hesperidin, is found as a major flavanone in peels of c*itrus unshiu* [23].

3.1. Extraction of Narirutin and Hesperidin

The extraction of bioactive compounds from citrus peel is the first crucial step in the utilization of flavonoids in the preparation of dietary supplements or food ingredients, and cosmetic products. Citrus flavonoids can be extracted from low-grade and dried plant materials or byproducts.

Citrus flavonoids are commonly extracted using aqueous-based ethanol solutions. Ethanol has a relatively low environmental impact because of recycling, and is classified as a 'generally recognized as safe' (GRAS) solvent, therefore it has often been used as an extraction solvent for food materials. In addition, subcritical water could be an excellent alternative medium for extracting flavonoids due to its temperature-dependent selectivity, environmental acceptability, and efficiency [24]. Subcritical condition is also referred to as high-temperature and pressurization [25]. However, these processes have many drawbacks, e.g., too much energy and time needed for extraction. To utilize citrus byproducts, energy saving and environment-friendly extraction procedures are recommended. Here, we briefly introduce application of the ultrasound-assisted extraction (UAE) to extract narirutin and hesperidin from peels of *C. unshiu*.

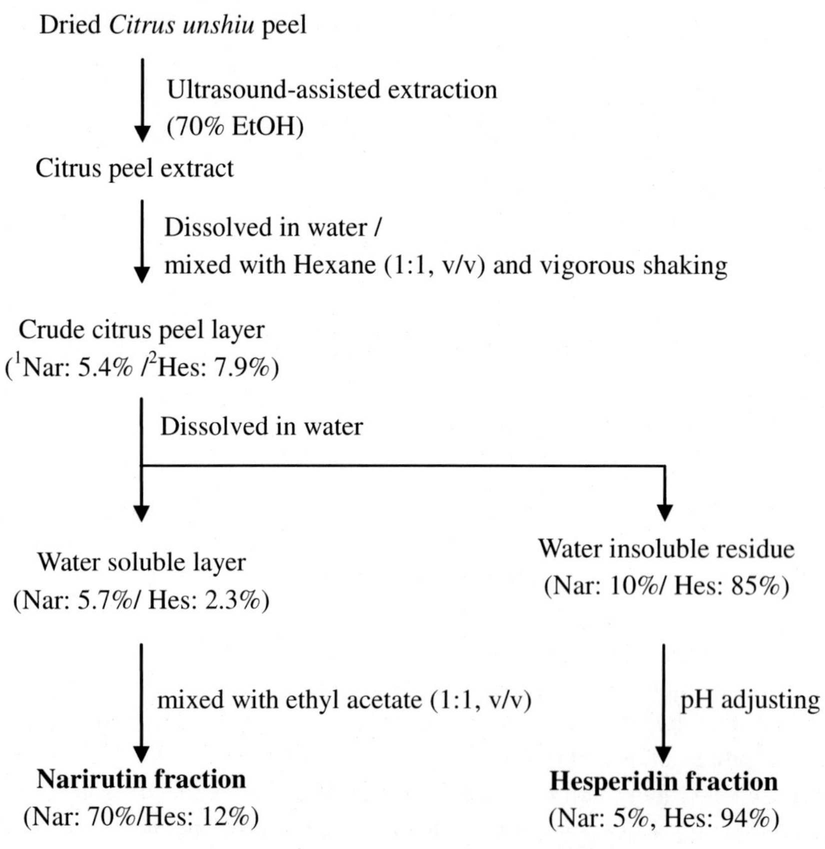

Figure 2. Isolation and purification procedure of narirutin and hesperidin fractions from dried peels of *Citrus unshiu*. [1]Nar, narirutin; [2]Hes, hesperidin.

UAE has advantages due to its higher extraction efficiency with shorter extraction time compared to conventional extracting methods. It uses sound waves at frequencies above 20 kHz to disrupt the plant cell wall, thereby enhancing solvent penetration into the plant material and improving the release of extracts [26]. In recent years, there have been several reports on the application of UAE in the isolation of various biologically active compounds from plant materials. For example, extraction of phenolic compounds from citrus peel [27], anthocyanins from red raspberries [28], polysaccharides from mulberry fruits [29], and essential oils from *Kaempferia galangal* [30] are reported. In general, there was an increase in the yield of phenolic compounds with increasing maceration time, whereas UAE at too high temperature resulted in reduction in the yields of the phenolics [27].

Recently, we demonstrated the marked reduction in extraction time of citrus flavonoids from citrus peels by UAE using 70% ethanol at 750 W ultrasonic processor (20 kHz at 30°C) [31]. The contents of narirutin and hesperidin in the crude extract were 5.4% and 9.7%, respectively after 10 min treatment [31]. Since crude citrus peel extracts usually contain large amounts of carbohydrates, fat-soluble materials (carotenoids and xanthophylls), and ashes, strategies including liquid-liquid partitioning (using *n*-hexane) and sequential separation based on polarity and water solubility have been used for further concentration of narirutin and hesperidin. To remove fat-soluble materials from crude extract, the equivalent amount of hexane was added, followed by vigorous stirring. Alkaliphilic and hydrophobic hesperidin were then extracted with alkaline treatments (pH 10–13) from citrus peels, and subsequently purified by the crystallization with the acidification (pH 5–7) of the alkaline extract [32, 33]. In brief, dried citrus extracts were resuspended in distilled water and separated into water-soluble and water-insoluble layers (Figure 2). The water insoluble residue was resuspended in distilled water, and 1 N NaOH was added until the pH reached 12 to make hesperidin soluble, followed by neutralization to pH 6 by adding 1 N HCl. The crystallized hesperidin was collected and lyophilized to have 94.2% purity. The water-soluble layer was mixed with same volume of ethyl acetate, and the ethyl acetate layer was collected and lyophilized to have 70.4% purity [34].

3.2. Physiological Properties of Narirutin and Hesperidin

3.2.1. Anti-Inflammatory Activities of Narirutin and Hesperidin

Flavonoids in citrus have long been used for their anti-inflammatory potential as either juices or dried peel extracts. In some cases, inflammation is a harmful biological process that involves the overproduction of various pro-inflammatory mediators or cytokines for protecting the host body from foreign molecules and pathogenic substance. For example, chronic inflammation is usually induced by overexpression of nitric oxide synthase (iNOS) and cyclooxygenase-2 (COX-2) as proinflammatory mediators, which increase the level of nitric oxide (NO) and prostaglandins (PGs), respectively. Moreover, increased levels of proinflammatory cytokines such as TNF-α, interleukin (IL)-1β, and IL-6 in serum can be risk factors for several chronic inflammatory diseases.

NF-κB is one of the major transcription factors that play a primary role in the expression of inflammation by inducing enzymes (iNOS and COX-2) and cytokines (TNF-α, IL-1β, and IL-6). Another signaling pathway of NF-κB is related to activation of mitogen-activated protein kinases (MAPKs), which are intracellular serine/threonine-specific protein kinases. In addition, intercellular adhesion molecule-1 (ICAM-1) and vascular cell adhesion molecule-1 (VCAM-1) also assist in inflammatory signal transduction by regulation of leukocyte–endothelial cell interactions [35, 36].

Results of many studies proved the attenuating effect of citrus flavonoids against inflammatory reactions [37]. These in vitro and in vivo studies used inflammatory agents such as lipopolysaccharide [38], benzo(a)pyrene [39] and hydrogen peroxide [40].

In our study, the narirutin-rich fraction (70.4% purity) isolated from citrus peels through the UAE method exhibited the regulatory activity of inflammatory mediators managing acute or chronic inflammatory disease in mouse macrophage cell line RAW 264.7 [41]. Citrus narirutin fraction (CNF) inhibited the translocation of the p65 subunit of NF-κB from the cytosol to the nucleus by decreasing the IκB-α degradation. These results demonstrate that CNF suppressed the release of NO and prostaglandin E2 (PGE2) in lipopolysaccharide (LPS)-stimulated macrophages by regulating the expression of iNOS and COX-2. In addition, CNF inhibited the phosphorylation of MAPKs (p38, ERK, and JNK) without any change in total protein expression indicating that narirutin is effective in suppressing inflammation through the inhibition of MAPKs phosphorylation (Figure 3). In the pre-clinical animal model, inhibiting MAPKs pathway with diverse agents showed effective anti-inflammatory activities [42].

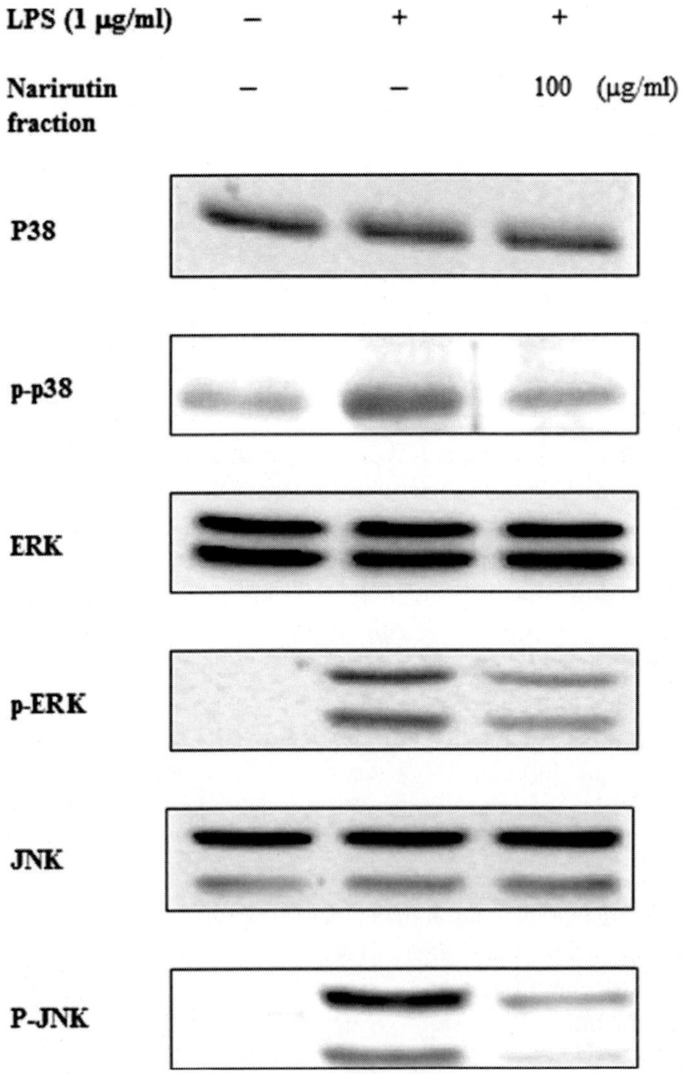

Figure 3. Effect of citrus narirutin fraction on LPS-induced MAPK activation (From Ha et al., *Food and Chemical Toxicology* 50, 2012, pp 3503).

Hesperidin also regulates the inflammatory response through inhibition of synthesis and release of proinflammatory mediators, inhibition of activated immune cells, or inhibiting such as COX-2, matrix metalloproteinase (MMP)-2, and MMP-9 [39, 43]. Several animal studies showed that hesperidin has preventive effects on inflammatory-associated neuropathology, nephrotoxicity, and cardiovascular complication. Pretreatment of hesperidin (100 and 200 mg/kg) to Swiss male albino mice prevented cognitive impairment induced by intracerebroventricular-streptozotocin (ICV-STZ) injection (2.57 mg/kg) through modulation of acetylcholine esterase activity (AChE), and modulates neuronal cell death by inhibiting the overexpression of inflammatory markers like NF-κB, iNOS, COX-2 and glial fibrillary acidic protein (GFAP) positive astrocytes [44]. In gentamicin-induced nephrotoxic rats, hesperidin (200 mg/kg) supplementation inhibited NF-κB mediated expression of inflammatory genes (IL-6, iNOS, TNF-α, and COX-2) and upregulated nuclear factor (erythroid-derived 2)-like 2 (Nrf2) and Nrf2 dependent phase II antioxidant gene HO-1 [45]. In addition, treatment of hesperidin (100 mg/kg) for 28 days prevented cardiovascular complication in isoproterenol-induced myocardial infarction in normal and streptozotocin-nicotinamide induced diabetic rats by decreasing HbA1c, glucose, CES, LDL, TC, TG, systolic levels, and diastolic blood pressure [46]. These results indicated that hesperidin attenuates the inflammatory response via direct transcriptional effect or decreased tissue lipid deposition.

3.2.2. *Preventive Effect of Narirutin and Hesperidin on Alcoholic Liver Disease*

Liver-related diseases are still one of the critical causes of disability and morbidity in the world. Fatty liver, an initial stage of alcoholic liver disease (ALD), is one of the most common liver diseases. Several reports have indicated that chronic alcohol consumption causes fatty liver or steatosis [47, 48], and may progress to hepatitis, inflammation, and fibrosis, which may eventually lead to liver cirrhosis [49].

Therapeutic agents have been successfully used for the prevention or treatment of major risk factors for ALD. Particularly, there is a growing interest in natural bioactive compounds that protect against or mitigate liver diseases without the undesirable side effects. Among those biocompounds, flavonoids are one of the best alternatives.

Our previous studies indicated that narirutin and hesperidin have potent anti-ALD activities. Their hepatoprotective effects were exhibited by maintaining antioxidant status, inhibiting lipid accumulation and apoptosis. Levels of hepatic triglyceride (TG) were significantly ($p < 0.05$) increased by ethanol-mediated oxidative stress, while such excessive formation of hepatic lipids was obviously inhibited by co-administration of citrus flavonoids. Investigations were carried out to determine whether administration of citrus flavonoids could regulate oxidative stress to inhibit excessive lipid formation in hepatocytes.

In biological systems, both enzymatic and non-enzymatic antioxidants including glutathione (GSH), glutathione S-transferase (GST), glutathione reductase (GR), glutathione peroxidase (GPx) and superoxide dismutase (SOD) play critical roles in handling oxidative stress [50]. It is normally observed that levels of these antioxidant components in hepatocytes are decreased by chronic and/or excessive alcohol consumption. Therefore, it is essential to maintain antioxidant systems to overcome alcohol exposure. Administration of narirutin could maintain these antioxidant systems. Co-administration of CNF (300 mg/kg mouse) reinstated the GSH levels (16.7 nmol/mg protein) and SOD activity (96.5% of normal group) to near normal controls [34]. Oral administration of citrus flavonoids (54% of hesperidin and 28% of narirutin) also prevented decrease of antioxidant enzyme activities including GR,

GST, and GPx in a significant level ($p < 0.05$) against ethanol consumption (5 g/kg mouse) for 8 weeks [33]. These results demonstrate that oral administration of narirutin and/or citrus flavonoids may suppress alcoholic fatty liver, via the antioxidant defense mechanism against ethanol-induced oxidative stress.

To maintain antioxidant systems is not only important for suppressing steatosis but also suppressing further damage to hepatocytes, which is caused by lipid peroxidation. Lipid peroxidation of hepatic tissues caused by ROS is another main cause of liver damage induced by ethanol and can produce harmful radicals [51, 52]. Setshedi et al. have also demonstrated that mouse liver produced large amounts of malondialdehyde (MDA) and 4-hydroxynonenal (4-HNE) mediated by CYP2E1, which resulted in increased generation of ROS leading to oxidative damage and cell death after excessive ethanol administration [53]. Likewise, chronic ethanol consumption caused an increase in MDA concentration in hepatocytes, which could be effectively inhibited by administration of citrus flavonoids.

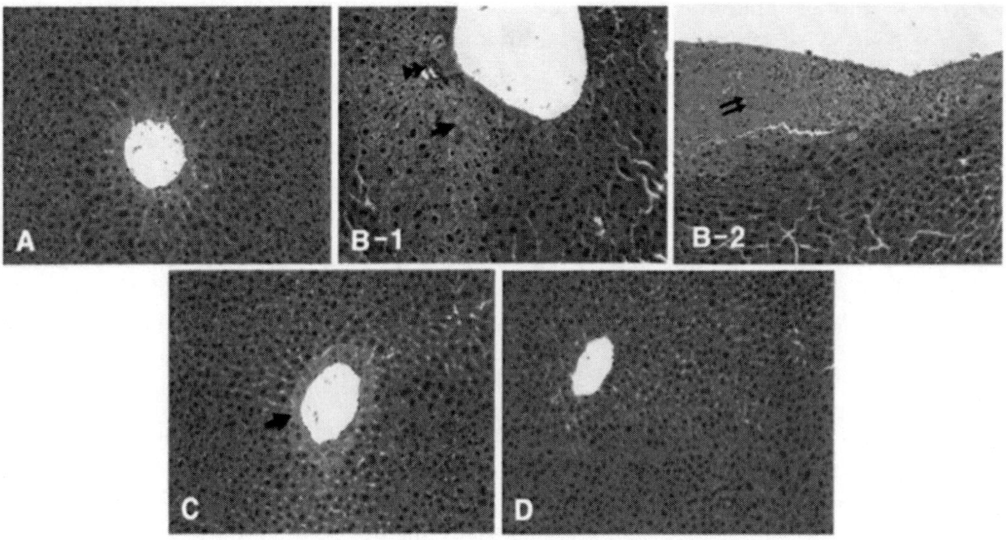

Figure 4. Histopathological changes of hepatic tissues after co-administration of citrus fraction along with 6.5 g ethanol/kg body weight/day for 8 weeks in ICR mice. No histological abnormalities were observed in normal mice (A). Mice induced with ethanol developed typical lesions of liver damage, which were increased in hydropic degeneration (arrows) and lymphocytes (double arrow head) around the portal vein (B-1) and occurrence of necrosis (double arrow) in hepatocytes (B-2). Co-administration of citrus narirutin fraction (150 mg/kg) resulted in suppression of necrosis and alleviation of increasing lymphocytes (C). Especially, 300 mg/kg of citrus narirutin fraction almost blocked the typical lesions of liver damage (D). The tissues were subjected to histological study by staining with hematoxylin and eosin (H&E). All the magnification is 200 × (From Park et al., *Food and Chemical Toxicology* 55, 2013, pp 642).

Liver inflammation is closely associated with ROS released from lipid peroxidation. As in the cases of other organs, activated NF-κB, a key transcription factor in inflammation, stimulates many inflammatory cytokine genes to trigger mediators in hepatocytes such as TNF-α, ILs, inducible nitric oxide synthase (iNOS), that are very important for ALD [54]. It has been demonstrated that NF-κB is activated by ROS (Schreck et al., 1991) and its activation can be modulated by antioxidants (Omoya et al., 2001). Narirutin can suppress levels of these inflammatory components through potent antioxidant [55] and anti-inflammatory effects [41].

Moreover, focal hepatocytic necrosis with lymphocytes infiltration observed in ethanol-induced mice might be a basis for cellular damage [56]. Administration of CNF (300 mg/kg) remarkably reduced the histological alterations caused by ethanol, which may be due to attenuation of the ethanol-mediated threat and reduction of the pathological changes (Figure 4). Chronic alcohol exposure showed that hepatocytes near the central vein area were full of hydropic degeneration, and infiltration of lymphocytes and appreciable necrosis were observed in the central vein area. On the other hand, liver plate and hepatocytes structure of CNF administered group were almost intact and the boundary between hepatocytes was clear.

3.3. Enzymatic Modification of Narirutin and Hesperidin

. In recent decades, isolation of plant flavonoids and their enzymatic-modified forms has become the subject of considerable scientific interest. Although flavonoids exhibit biological benefits and are demanded in functional foods and pharmaceuticals, they have some drawbacks such as poor water solubility and low bioavailability that could inhibit flavonoids from exercising their physiological properties fully in a body. Enzymatic modifications like addition or cleavage of sugars on flavonoids suggest one of the best alternatives to overcome such drawbacks. In general, health related properties of flavonoids and their glycosylated or deglucosylated forms differ significantly. Detailed knowledge of these differences is an essential prerequisite for the use of bioconversion in the production of modified natural substances, which have the potential to prevent many serious diseases [57]. Enzymatic modification of flavonoids enhanced their own biological impacts including antioxidant and anticancer properties. The use of enzymes to modify the structure of flavonoids is of industrial interest due to their high selectivity, low cost, and wide availability.

Glycosylation is a popular modification reaction of natural compounds, which result in the formation of natural glucosides. The reaction converts starch into cyclic α-1,4-linked oligosaccharides (cyclodextrins, CD) using cyclodextrin glucanotransferases (CGTases; EC 2.4.1.19). CGTases, certain extracellular bacterial enzymes, involve intracellular transglycosylation that monopolize on the starch substrate, converting into CDs. CGTase cleave the α-1,4-glycosidic bonds between the subsites in α-glucans yielding a stable covalent glycosyl-intermediate bond at the donor subsites, and then the glycosyl-intermediate is then transferred to its own non-reducing end, forming a new α-1,4-glycosidic bond with flavonoids [58]. In a structural analysis of glycosylated naringin, both the hydroxyl group of the glucose moiety and the phenolic group in naringin were subjected to the glycosylation, and CGTase mainly glucosylated the 3″-hydroxyl group of the glucose moiety [59]. Poor water-solubility of hesperidin and naringenin were enhanced by complexation with glucosides at their hydrophilic outside. The solubility of both hesperidin mono- and di-glucoside in water was about 300 times higher than that of hesperidin [60]. In the presence of β-CD, however, the solubility of naringenin increased up to 500-fold [61].

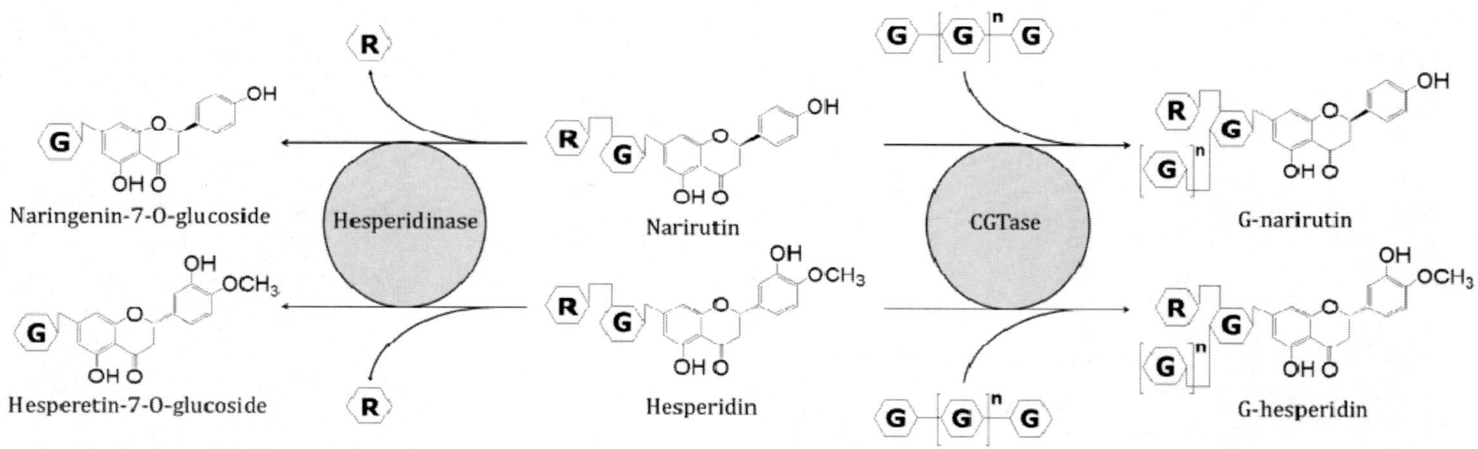

Figure 5. Enzymatic bioconversion of citrus narirutin and hesperidin. Narirutin was converted to glycosylated narirutin (G-narirutin) and naringenin-7-*O*-glucoside using CGTase and hesperidinase, respectively. Hesperidin fraction was also converted to glycosylated hesperidin (G-hesperidin) and hesperetin-7-*O*-glucoside.

Previous work has shown that enzymatic hydrolysis of specific glycosyl groups or the conversion of flavonoid glycosides to aglycones increases antioxidant activity of rutin [62], anti-inflammatory activity of naringin [63], antioxidant activity of kaempferol [64], and improves the bioavailability of hesperidin [65]. CDs are also used to suppress undesirable tastes and improve the stability of ingredients such as flavors, vitamins, and unsaturated fats [66].

In our previous study (Figure 5), citrus flavonoids and their enzymatically modified ones were compared to evaluate the bioavailability in oral administered rat model [33]. Hesperidin was mainly metabolized to glucuronide conjugates in the intestinal tissue and liver after hydrolysis into hesperetin by intestinal microflora [67]. Glucuronide and sulfate conjugates of hesperetin in plasma were found more in rats administered hesperetin-7-glucoside and G-hesperidin than in those in the hesperidin administered group, which was indicating easier absorption of enzymatically modified hesperidin into a body than untreated hesperidin. The absorption rate of flavonoids is affected by its solubility in a digestion process [65]. Therefore, elevated absorption efficiency through glycosylation reaction is due to their increased water solubility. Less than 0.01% of hesperidin was soluble in water whereas water solubility was about 10,000 times increased by glycosylation [68].

On the other hand, rutinoside of flavonoids require de-glycosylation of rhamose for absorption across the intestine [69]. Although, glycans of terminal rhamnose are not hydrolyzed in human intestinal tissue, certain strains of bacteria were able to hydrolyze a rhamosyl moiety (de-rhamnosylation from rutinosides) of flavonoids [65]. Flavonoids with mono-glucoside are suitable for absorption at the small intestine instead of the colon and are transported into the enterocytes via a human intestinal sugar transporter system [70] and UDP-glucuronosyltransferases [71]. Removal of rutinoside from narirutin and hesperidin enhanced their physiological activities through their increased absorption into a body [72].

CONCLUSION

Dietary supplementation of citrus fruits seemed like an appropriate route for improving health. Citrus fruits are rich in innumerable chemical compounds including vitamins, flavonoids, minerals, and other biologically active compounds. Furthermore, the consumption of a diet abundant in citrus peel extract typically contains high amount of flavonoids.

A number of therapeutically relevant compounds are derived from citrus fruits and its byproducts. Therefore, evaluation of citrus flavonoids as metabolic regulators represents an established avenue for drug discovery. In this chapter, it is reviewed that citrus flavonoids possess significant anti-oxidative and anti-inflammatory potential, and demonstrate that these compounds, particularly narirutin and hesperidin, prevent alcohol-induced liver damage.

Many studies give evidence that anti-inflammatory action of flavonoids can provide various benefits to human health. Exactly which mechanisms operate in which circumstances still requires further illumination. Further studies are essential to fully reveal the interaction of these flavonoids with downstream mediators of pathways. In addition, clinical studies of dosage adjustment, ADME and safety may be further required for application of citrus nariautin and hesperidin to nutraceuticals for liver health.

REFERENCES

[1] Henderson AJ, Kumar A, Barnett B, Dow SW, Ryan EP. (2012). Consumption of rice bran increases mucosal immunoglobulin A concentrations and numbers of intestinal Lactobacillus spp. *J Med Food.*

[2] Barros HRdM, Ferreira TAPdC, Genovese MI. (2012). Antioxidant capacity and mineral content of pulp and peel from commercial cultivars of citrus from Brazil. *Food Chem*, 134:1892-8.

[3] Ladaniya MS. 17 - Physiological disorders and their management. In: Ladaniya MS, editor. Citrus Fruit. San Diego: Academic Press; 2008. p. 451-63.

[4] Tsao R. (2010). Chemistry and biochemistry of dietary polyphenols. *Nutrients*, 2:1231-46.

[5] Neuhouser ML. (2004). Dietary flavonoids and cancer risk: evidence from human population studies. *Nutr Cancer*, 50:1-7.

[6] Busch C, Burkard M, Leischner C, Lauer UM, Frank J, Venturelli S. (2015). Epigenetic activities of flavonoids in the prevention and treatment of cancer. *Clin Epigenetics*, 7:64.

[7] Welch AA, Hardcastle AC. (2014). The effects of flavonoids on bone. *Curr Osteoporos Rep*, 12:205-10.

[8] McCullough ML, Peterson JJ, Patel R, Jacques PF, Shah R, Dwyer JT. (2012). Flavonoid intake and cardiovascular disease mortality in a prospective cohort of US adults. *Am J Clin Nutr*, 95:454-64.

[9] Jennings A, Welch AA, Spector T, Macgregor A, Cassidy A. (2014). Intakes of anthocyanins and flavones are associated with biomarkers of insulin resistance and inflammation in women. *J Nutr*, 144:202-8.

[10] Solanki I, Parihar P, Mansuri ML, Parihar MS. (2015). Flavonoid-based therapies in the early management of neurodegenerative diseases. *Adv Nutr*, 6:64-72.

[11] Mouly PP, Arzouyan CR, Gaydou EM, Estienne JM. (1994). Differentiation of citrus juices by factorial discriminant analysis using liquid chromatography of flavanone glycosides. *J Agric Food Chem*, 42:70-9.

[12] Li C, Schluesener H. (2015). Health-promoting effects of the citrus flavanone hesperidin. *Crit Rev Food Sci Nutr*:0.

[13] Harborne JB. (1999). The comparative biochemistry of phytoalexin induction in plants. *Biochem Syst Ecol*, 27:335-67.

[14] USDA database for the flavonoid content of selected fruits, release 2.1. U.S. Department of Agriculture; 2007.

[15] Choudhury R, Chowrimootoo G, Srai K, Debnam E, Rice-Evans CA. (1999). Interactions of the flavonoid naringenin in the gastrointestinal tract and the influence of glycosylation. *Biochem Biophys Res Commun*, 265:410-5.

[16] Hsiu S-L, Huang T-Y, Hou Y-C, Chin D-H, Chao P-DL. (2002). Comparison of metabolic pharmacokinetics of naringin and naringenin in rabbits. *Life Sci*, 70:1481-9.

[17] Cavia-Saiz M, Busto MD, Pilar-Izquierdo MC, Ortega N, Perez-Mateos M, Muniz P. (2010). Antioxidant properties, radical scavenging activity and biomolecule protection capacity of flavonoid naringenin and its glycoside naringin: a comparative study. *J Sci Food Agric*, 90:1238-44.

[18] Ribeiro IA, Rocha J, Sepodes B, Mota-Filipe H, Ribeiro MH. (2008). Effect of naringin enzymatic hydrolysis towards naringenin on the anti-inflammatory activity of both compounds. *J Mol Catal B: Enzym*, 52–53:13-8.

[19] Alam MA, Subhan N, Rahman MM, Uddin SJ, Reza HM, Sarker SD. (2014). Effect of citrus flavonoids, naringin and naringenin, on metabolic syndrome and their mechanisms of action. *Adv Nutr*, 5:404-17.

[20] Lee S, Ra J, Song JY, Gwak C, Kwon HJ, Yim SV, et al. (2011). Extracts from *Citrus unshiu* promote immune-mediated inhibition of tumor growth in a murine renal cell carcinoma model. *J Ethnopharmacol*, 133:973-9.

[21] Jung UJ, Lee MK, Park YB, Kang MA, Choi MS. (2006). Effect of citrus flavonoids on lipid metabolism and glucose-regulating enzyme mRNA levels in type-2 diabetic mice. *Int J Biochem Cell Biol*, 38:1134-45.

[22] But PPH, Kimura T, Guo J-X, Sung CK. *International collation of traditional and folk medicine: Northeast Asia*. River Edge: World Scientific Publishing Co.; 1996.

[23] Whitman SC, Kurowska EM, Manthey JA, Daugherty A. (2005). Nobiletin, a citrus flavonoid isolated from tangerines, selectively inhibits class A scavenger receptor-mediated metabolism of acetylated LDL by mouse macrophages. *Atheroscler*, 178:25-32.

[24] Cheigh C-I, Chung E-Y, Chung M-S. (2012). Enhanced extraction of flavanones hesperidin and narirutin from *Citrus unshiu* peel using subcritical water. *J Food Eng*, 110:472-7.

[25] Aliakbarian B, Fathi A, Perego P, Dehghani F. (2012). Extraction of antioxidants from winery wastes using subcritical water. *J Supercrit Fluids*, 65:18-24.

[26] Mason TJ, Paniwnyk L, Lorimer JP. (1996). The uses of ultrasound in food technology. *Ultrason Sonochem*, 3:S253-S60.

[27] Karg SR, Frey AD, Ferrara C, Streich DK, Umana P, Kallio PT. (2009). A small-scale method for the preparation of plant N-linked glycans from soluble proteins for analysis by MALDI-TOF mass spectrometry. *Plant Physiol Biochem*, 47:160-6.

[28] Chen F, Sun Y, Zhao G, Liao X, Hu X, Wu J, et al. (2007). Optimization of ultrasound-assisted extraction of anthocyanins in red raspberries and identification of anthocyanins in extract using high-performance liquid chromatography–mass spectrometry. *Ultrason Sonochem*, 14:767-78.

[29] Chen C, You L-J, Abbasi AM, Fu X, Liu RH. (2015). Optimization for ultrasound extraction of polysaccharides from mulberry fruits with antioxidant and hyperglycemic activity in vitro. *Carbohydr Polym*, 130:122-32.

[30] Ma Q, Fan X-D, Liu X-C, Qiu T-Q, Jiang J-G. (2015). Ultrasound-enhanced subcritical water extraction of essential oils from *Kaempferia galangal* L. and their comparative antioxidant activities. *Sep Purif Technol*, 150:73-9.

[31] Park H-Y, Park Y, Lee Y, Noh SK, Sung E-G, Choi I. (2012). Effect of oral administration of water-soluble extract from citrus peel (*Citrus unshiu*) on suppressing alcohol-induced fatty liver in rats. *Food Chem*, 130:598-604.

[32] Kim M-R, Kim W-C, Lee D-Y, Kim C-W. (2007). Recovery of narirutin by adsorption on a non-ionic polar resin from a water-extract of Citrus unshiu peels. *J Food Eng*, 78:27-32.

[33] Park H-Y, Choi H-D, Eom H, Choi I. (2013). Enzymatic modification enhances the protective activity of citrus flavonoids against alcohol-induced liver disease. *Food Chem*, 139:231-40.

[34] Park H-Y, Ha SK, Eom H, Choi I. (2013). Narirutin fraction from citrus peels attenuates alcoholic liver disease in mice. *Food Chem Toxicol*, 55:637-44.

[35] Cook-Mills JM, Marchese ME, Abdala-Valencia H. (2011). Vascular cell adhesion molecule-1 expression and signaling during disease: regulation by reactive oxygen species and antioxidants. *Antioxid Redox Signal*, 15:1607-38.

[36] Hubbard AK, Rothlein R. (2000). Intercellular adhesion molecule-1 (ICAM-1) expression and cell signaling cascades. *Free Radic Biol Med*, 28:1379-86.

[37] Benavente-Garcia O, Castillo J. (2008). Update on uses and properties of citrus flavonoids: new findings in anticancer, cardiovascular, and anti-inflammatory activity. *J Agric Food Chem*, 56:6185-205.

[38] Xie C, Kang J, Li Z, Schauss AG, Badger TM, Nagarajan S, et al. (2012). The acai flavonoid velutin is a potent anti-inflammatory agent: blockade of LPS-mediated TNF-alpha and IL-6 production through inhibiting NF-kappaB activation and MAPK pathway. *J Nutr Biochem*, 23:1184-91.

[39] Kamaraj S, Anandakumar P, Jagan S, Ramakrishnan G, Devaki T. (2010). Modulatory effect of hesperidin on benzo(a)pyrene induced experimental lung carcinogenesis with reference to COX-2, MMP-2 and MMP-9. *Eur J Pharmacol*, 649:320-7.

[40] Wang A-Y, Zhou M-Y, Lin W-C. (2011). Antioxidative and anti-inflammatory properties of *Citrus sulcata* extracts. *Food Chem*, 124:958-63.

[41] Ha SK, Park H-Y, Eom H, Kim Y, Choi I. (2012). Narirutin fraction from citrus peels attenuates LPS-stimulated inflammatory response through inhibition of NF-κB and MAPKs activation. *Food Chem Toxicol*, 50:3498-504.

[42] Mayer RJ, Callahan JF. (2006). p38 MAP kinase inhibitors: A future therapy for inflammatory diseases. *Drug Discov Today*, 3:49-54.

[43] Yeh MH, Kao ST, Hung CM, Liu CJ, Lee KH, Yeh CC. (2009). Hesperidin inhibited acetaldehyde-induced matrix metalloproteinase-9 gene expression in human hepatocellular carcinoma cells. *Toxicol Lett*, 184:204-10.

[44] Javed H, Vaibhav K, Ahmed ME, Khan A, Tabassum R, Islam F, et al. (2015). Effect of hesperidin on neurobehavioral, neuroinflammation, oxidative stress and lipid alteration in intracerebroventricular streptozotocin induced cognitive impairment in mice. *J Neurol Sci*, 348:51-9.

[45] Subramanian P, Anandan R, Jayapalan JJ, Hashim OH. (2015). Hesperidin protects gentamicin-induced nephrotoxicity via Nrf2/HO-1 signaling and inhibits inflammation mediated by NF-κB in rats. *J Funct Foods*, 13:89-99.

[46] Kakadiya J, Mulani H, Shah N. (2010). Protective effect of hesperidin on cardiovascular complication in experimentally induced myocardial infarction in diabetes in rats. *J Basic Clin Pharm*, 1:85-91.

[47] Brandon-Warner E, Schrum LW, Schmidt CM, McKillop IH. (2012). Rodent models of alcoholic liver disease: Of mice and men. *Alcohol*, 46:715-25.

[48] Lieber CS. (2004). Alcoholic fatty liver: its pathogenesis and mechanism of progression to inflammation and fibrosis. *Alcohol*, 34:9-19.

[49] Sharma R. 41 - Biochemical mechanisms of fatty liver and bioactive foods: Fatty Liver, diagnosis, nutrition therapy. In: Preedy RRWR, editor. Bioactive Food as Dietary

Interventions for Liver and Gastrointestinal Disease. San Diego: Academic Press; 2013. p. 623-55.

[50] Albano E. (2006). Alcohol, oxidative stress and free radical damage. *Proc Nutr Soc*, 65:278-90.

[51] Kashinakunti SV, Kallaganada GS, Manjula Rangappa, Hiremath K, Bannale S. (2012). Are we over estimating serum oxidative stress/lipid peroxidation in alcoholic liver diseases? A review. *Free Radicals and Antioxidants*, 2:1-9.

[52] Crosas-Molist E, Fabregat I. (2015). Role of NADPH oxidases in the redox biology of liver fibrosis. *Redox Biol*, 6:106-11.

[53] Setshedi M, Wands JR, de la Monte SM. (2010). Acetaldehyde adducts in alcoholic liver disease. *Oxid Med Cell Longev*, 3:178-85.

[54] Song Z, Joshi-Barve S, Barve S, McClain CJ. (2004). Advances in alcoholic liver disease. *Curr Gastroenterol Rep*, 6:71-6.

[55] Tripoli E, Guardia ML, Giammanco S, Majo DD, Giammanco M. (2007). Citrus flavonoids: Molecular structure, biological activity and nutritional properties: A review. *Food Chem*, 104:466-79.

[56] Brooks PJ. (1997). DNA damage, DNA repair, and alcohol toxicity-a review. *Alcohol Clin Exp Res*, 21:1073-82.

[57] Karabín M, Hudcová T, Jelínek L, Dostálek P. (2015). Biotransformations and biological activities of hop flavonoids. *Biotechnol Adv*, 1:1063-90.

[58] Leemhuis H, Kelly RM, Dijkhuizen L. (2010). Engineering of cyclodextrin glucanotransferases and the impact for biotechnological applications. *Appl Microbiol Biotechnol*, 85:823-35.

[59] Akiyama T, Yamada M, Yamada T, Maitani T. (2000). Naringin glycosides alpha-glucosylated on ring B found in the natural food additive, enzymatically modified naringin. *Biosci Biotechnol Biochem*, 64:2246-9.

[60] Kometani T, Terada Y, Nishimura T, Takii H, Okada S. (1994). Transglycosylation to hesperidin by cyclodextrin glucanotransferase from an alkalophilic Bacillus species in alkaline pH and properties of hesperidin glycosides. *Biosci Biotechnol Biochem*, 58:1990-4.

[61] Shulman M, Cohen M, Soto-Gutierrez A, Yagi H, Wang H, Goldwasser J, et al. (2011). Enhancement of naringenin bioavailability by complexation with hydroxypropyl-beta-cyclodextrin. *PLoS One*, 6:e18033.

[62] de Araújo MEMB, Moreira Franco YE, Alberto TG, Sobreiro MA, Conrado MA, Priolli DG, et al. (2013). Enzymatic de-glycosylation of rutin improves its antioxidant and antiproliferative activities. *Food Chemistry*, 141:266-73.

[63] Amaro M, Rocha J, Vila-Real H, Eduardo-Figueira M, Mota-Filipe H, Sepodes B, et al. (2009). Anti-inflammatory activity of naringin and the biosynthesised naringenin by naringinase immobilized in micro-structured materials in a model of DSS-induced colitis in mice. *Food Res Int*, 42:1010-7.

[64] Park JS, Rho HS, Kim DH, Chang IS. (2006). Enzymatic preparation of kaempferol from green tea seed and its antioxidant activity. *J Agric Food Chem*, 54:2951-6.

[65] Nielsen IL, Chee WS, Poulsen L, Offord-Cavin E, Rasmussen SE, Frederiksen H, et al. (2006). Bioavailability is improved by enzymatic modification of the citrus flavonoid hesperidin in humans: a randomized, double-blind, crossover trial. *J Nutr*, 136:404-8.

[66] Szente L, Szejtli J. (2004). Cyclodextrins as food ingredients. *Trends Food Sci Tech*, 15:137-42.

[67] Matsumoto H, Ikoma Y, Sugiura M, Yano M, Hasegawa Y. (2004). Identification and quantification of the conjugated metabolites derived from orally administered hesperidin in rat plasma. *J Agric Food Chem*, 52:6653-9.

[68] Yamada M, Tanabe F, Arai N, Mitsuzumi H, Miwa Y, Kubota M, et al. (2006). Bioavailability of glucosyl hesperidin in rats. *Biosci Biotechnol Biochem*, 70:1386-94.

[69] Nemeth K, Plumb GW, Berrin JG, Juge N, Jacob R, Naim HY, et al. (2003). Deglycosylation by small intestinal epithelial cell beta-glucosidases is a critical step in the absorption and metabolism of dietary flavonoid glycosides in humans. *Eur J Nutr*, 42:29-42.

[70] Day AJ, Gee JM, DuPont MS, Johnson IT, Williamson G. (2003). Absorption of quercetin-3-glucoside and quercetin-4'-glucoside in the rat small intestine: the role of lactase phlorizin hydrolase and the sodium-dependent glucose transporter. *Biochem Pharmacol*, 65:1199-206.

[71] Walle UK, Walle T. (2002). Induction of human UDP-glucuro-nosyltransferase UGT1A1 by flavonoids-structural requirements. *Drug Metab Dispos*, 30:564-9.

[72] Manach C, Morand C, Gil-Izquierdo A, Bouteloup-Demange C, Remesy C. (2003). Bioavailability in humans of the flavanones hesperidin and narirutin after the ingestion of two doses of orange juice. *Eur J Clin Nutr*, 57:235-42.

In: Citrus Fruits
Editor: Daphne Simmons

ISBN: 978-1-63484-078-1

Chapter 4

CITRUS RESIDUES AS RAW MATERIALS FOR BIOMOLECULES AND ENERGY: THE ORANGE PEEL CASE

Valentina Hernández, Laura V. Daza and Carlos A. Cardona*

Instituto de Biotecnología y Agroindustria,
Universidad Nacional de Colombia Sede Manizales, Manizales, Colombia

ABSTRACT

The replacement of the fossil-based raw materials either fully or partially is an objective in many countries, being of special interest the use of local biomass such as agricultural, forest, agro-industrial and industrial wastes, due to its low cost and large availability. According to FAOSTAT, by 2011 approximately 120 million tons of citrus were produced worldwide, with oranges accounting approximately 63.1 million tons. Approximately 60% of the total citrus production is market for fresh consumption, while the other 40% is used in the agroindustry to extract no more than the 50% of the fruit weight as juice. Residues from agroindustrial processing are composed by peel, seeds and remaining pulp and, in most of the cases, are used to spread soils, to produce animal feed, or to be burned. However, these conventional disposal methods can cause negative effects on the soil and superficial waters. Moreover, several value-added products, such as phytochemicals, pharmaceuticals, food products, essential oils, seed oil, pectin and dietary fibers, can be obtained from orange residues. In this chapter, simulation results of the production of biofertilizers, gibberellic acid and electricity from orange peel as stand-alone products are presented. Moreover, the experimental characterization was assessed. Results from the characterization procedures have been used to feed the simulations to obtain the mass and energy balances that were subsequently used to perform the economic and environmental analysis of the above mentioned processes. Moreover, comparisons from the techno-economic and environmental points of view of the stand-alone processes were performed. Besides, and based on the experimental results of the

* Corresponding author: Tel: +57 6 8879400x55880; E-mail address: ccardonaal@unal.edu.co (Carlos A. Cardona).

physicochemical characterization, two biorefinery schemes were techno-economic and environmentally evaluated.

Keywords: citrus residues, biorefinery, limonene, pectin, biofertilizers, gibberellic acid, techno economic and environmental assessment

1. INTRODUCTION

Citrus is the most widely produced fruit in the world and it is cultivated in more than 80 countries. According to FAOSTAT, by 2011 approximately 120 million tons of citrus were produced worldwide, with oranges accounting approximately 63.1 million tons in 2010. The top five producing countries in 2013 (FAOSTAT) were Brazil, producing approximately 1/3 of the world production, followed by United States, China, India and Mexico. Brazilian citrus production is oriented toward processing, while USA citrus production is focused toward processing and the fresh fruit market. In Colombia the citrus are the most economically important fruit after banana. Although, nowadays Colombia is a smallholder producer compared to Brazil and USA, nevertheless many expansion possibilities appear in the west zones of the country. By 2011, Colombia produced approximately 207 thousand tons of orange coming from more than 31 thousand cultivated hectares, which implies a yield of 11 tons/ha [1]. Antioquia is the highest producer in the country (See Figure 1).

Approximately 60% of the total citrus production is consumed as fresh fruit while the remaining 40% is used in the agroindustry to extract no more than the 50% of the fruit weight as juice [3], while the peel, seeds and remaining pulp are considered wastes. This implies that citrus residues correspond to approximately the 20% of the world citrus production (24 million tons). The main products obtained from orange processing are juice, jam, essential oils, flavonoids and pectin. During these production processes, only approximately the 50% of the fruit is utilized [4-6]. The wastes generated and composed of peel, seeds and membrane residues are used to spread on soils, to produce animal feed, or to be burnt [5].

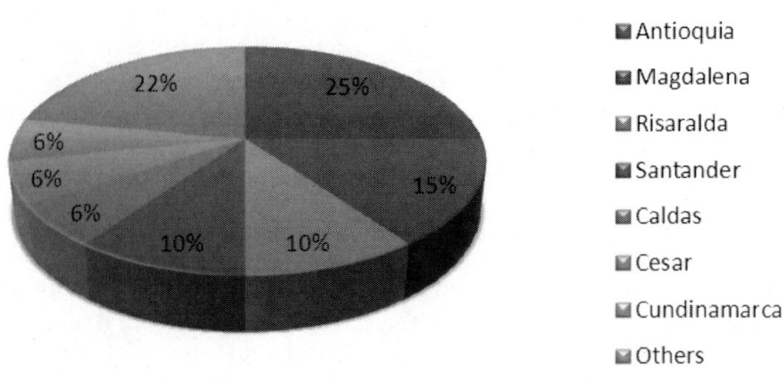

Figure 1. Colombian share production of citrus by MADR 2011 [2].

In this chapter, simulation results of the production of biofertilizers, gibberellic acid and electricity from orange peel (OP) as stand-alone products are presented. Moreover, the experimental characterization was assessed. Results from the characterization procedures have been used to feed the simulations to obtain the mass and energy balances that were subsequently used to perform the economic and environmental analysis of the above mentioned processes. Moreover, comparisons from the techno-economic and environmental points of view of the stand-alone processes were performed. Besides, and based on the experimental characterization, two biorefinery schemes were techno-economic and environmentally evaluated.

2. Integral Use of Lignocellulosic Waste

Currently, most energy and chemicals are derived from fossil raw materials, although there are several environmental, economic and social concerns related to their extraction and use. These facts have increased the interest in using renewable raw materials. Lignocellulosic biomass represents the major renewable source of potentially fermentable carbohydrates and it is mainly composed of cellulose, hemicellulose and lignin. The lignocellulosic complex constitute the principal component of most agricultural, forestry and municipal solid waste and byproducts from the agroindustry, food industry and other industrial sectors [7]. Its composition, availability and renewability have increased the scientific and academic interest in evaluating the production and recovery of a wide range of value-added products, such as enzymes, reducing sugars, furfural, biofuels, proteins and amino acids, lipids, organic acids, phenols, activated carbon, biopolymers, cosmetics, biosorbents, resins, medicines, bioenergy, biopesticides, biofertilizers, secondary metabolites, biopromotores and surfactants, among others [7-23].

Depending on factors such as residue type and amount and the final use of the product obtained by the biomass processing, a wide range of conversion technologies are available. These are classified into thermochemical, biochemical and chemical conversions as shown in Figure 2.

Thermochemical processes involve irreversible chemical reactions at high temperatures and different oxidation conditions. Into this classification, pyrolysis, gasification and combustion are included and they differ not only in process parameters but also in the products. The biochemical conversion covers those processes in which the action of microorganisms or enzymes is involved in the production process. The biochemical processes are referred mainly to a) fermentation, where a wide range of products such as ethanol, xylitol, gibberellic acid and compost, among others are produced and b) anaerobic digestion where biofertilizers and biogas are obtained.

2.1. Biorefinery Concept

Analogically to the refinery concept, in a biorefinery a wide range of marketable bio-based products, e.g., food, biofuels, biomaterials and biochemicals, are obtained through the sustainable processing of the raw material [12, 13]. The main differences between a refinery and a biorefinery are the nature of the raw material (biomass for a biorefinery) and the possibility of applying different existing and emerging technologies to obtain the bioproducts. Furthermore, in a biorefinery not only the products obtained in an oil refinery, but also products that cannot be obtained from crude oil, are included.

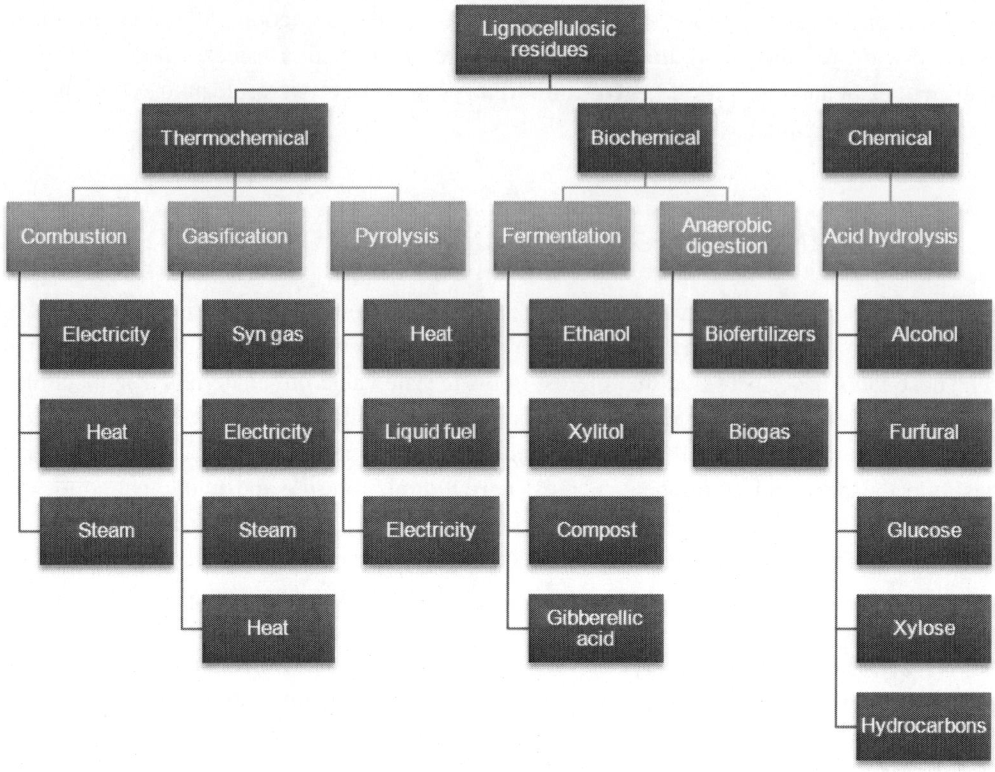

Figure 2. Conversion routes of the lignocellulosic biomass (Adapted from [24]).

Biorefining embraces to assess and use a broad range of technologies to separate the biomass into its principal constituents e.g., carbohydrates, proteins, oils, to be subsequently transformed into value-added products. A biorefinery comprises three main process operations: The first stage comprises the raw material supply. Second and third stages correspond to separation and conversion, respectively. For the first stage, agriculture, forest and agroindustrial materials as well as waste streams can be considered as feedstocks. The separation stage is generally referred to the pretreatment of the raw material to generate starch, cellulose, hemicellulose and monomeric sugars, as carbohydrates sources; lignin as aromatics source; and, hydrocarbons in the form of triglycerides. After the pretreatment, the biomass components are subjected to a combination of conversion technologies to obtain a bio-based family of specialty chemicals and fuels [25-32].

2.2. Feedstocks and Products

The biomass is synthesized by plants through the photosynthesis process, after both atmospheric carbon and water are converted into sugars. The main components of biomass are cellulose, hemicellulose and lignin. Cellulose represents between 40-50% of this complex, while hemicellulose and lignin represent 20-30% and 15-20%, respectively [33]. Other compounds such as proteins, terpenes, minerals, colorants, vitamins, flavors and lipids/oils, are also present in these feedstocks [30, 34]. Therefore, the aim of a biorefinery is the fractionation and further processing of the biomass in order to obtain value-added products with production volumes generally inversely proportional to its commercial values. In a biorefinery one or several low-volume and high-value chemical products are produced such as high-volume/low-value liquid fuel and low-volume/high-value specialty chemicals together, with power as electricity and process heat meeting national needs and reducing green house gas emissions [35-39].

Biomass resources to be used as raw materials in a biorefinery can be classified into first, second and third generation feedstocks. First generation feedstocks include edible crops such as sugarcane, oil palm, maize and fruits, among others. Non-edible energy crops, e.g., Jatropha, as well as crop residues, sustainably-harvested wood and forest residues, and clean municipal and industrial wastes are considered as second generation feedstocks. Finally, microalgae are considered as three generation feedstock. Moncada et al. and Hernández et al. [11, 12] classified the products into the following six categories: biofuels, bioenergy (referred as direct energy), biomolecules and natural chemicals, biofertilizers, biomaterials and food products, as the Figure 3 shows.

Feedstocks included into the first generation are not only renewable, but also have feed/food uses. Although, these rely on the high productivity of the crops, the population increase has resulted in concerns about competition with food needs [8, 40]. Second generation feedstocks are not directly used as food, although some of them are used as livestock feed. Sccond generation feedstocks are more beneficial than first generation feedstocks in terms of efficient land use and proper environmental management. Furthermore, these raw materials are leading the way to sustainably meeting energy needs while also supplying materials for chemical and manufacturing industries. Besides, to use organic wastes as raw materials is especially beneficial: When these residues are disposed of in a landfill, it decomposes and releases methane, which is a potent global warming gas. Thus, processing these wastes to obtain value-added products reduces landfill volume and reduces methane emissions [40].

Third generation feedstocks are considered a viable resource without the disadvantages associated to the first and second generation feedstocks. Microalgae are single-celled photosynthetic organisms located in different environments, under a wide range of conditions such as temperature, pH and nutrient availability. Furthermore, microalgae have a very high photosynthetic efficiency, contributing to the sequestration of atmospheric carbon dioxide. In addition, the residues from the algae processing can be included into the second generation feedstocks (See Figure 3).

2.3. Processes

Four main groups of processes are applied in biorefineries: thermochemical, biochemical, mechanical and chemical. In order to convert the biomass into value-added products, different combinations of processes can be implemented. Mainly three processes are considered into the first group: Direct combustion, gasification and pyrolysis. In the direct combustion process the biomass is oxidized to obtain heat and electricity. The gasification process is carried out in low oxygen environments and at temperatures over 700°C, to obtain syngas, which can be either used directly for producing heat and electricity or further converted into chemicals or fuel. Finally the pyrolysis process proceeds in an oxygen free environment, at temperatures range between 300-600°C, converting the biomass into oil, charcoal and a gaseous phase, similar to the syngas. Currently, the oil and char obtained from this process can be used as fuel in stationary combustion operations [41]. However, using the pyrolysis oil as transportation fuel is still under research [42].

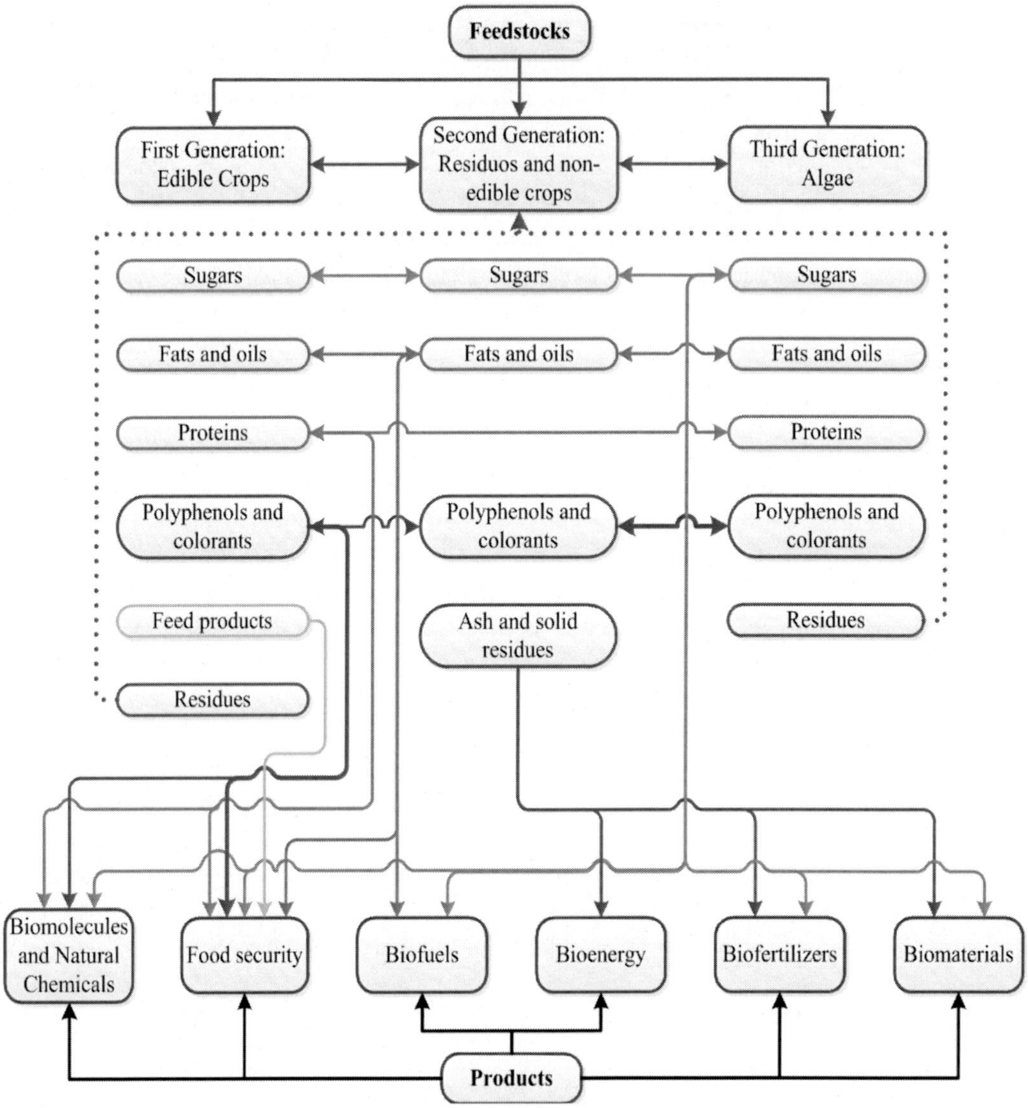

Figure 3. Feedstocks and products classification for biorefineries [11, 12].

Regarding to the biochemical route, two processes integrate this group: Anaerobic digestion (AD) and fermentation. In the case of the AD, two products are obtained: Biogas, which consist of methane and carbon dioxide and can be used directly or upgraded to biomethane by removing the carbon dioxide to be further either delivered to the natural gas grid or used as vehicle fuel [41]; and sludge with fertilizer qualities, which consists of a liquid and a solid fraction [43]. In the case of the fermentation process, the main products are alcohols and organic acids through using microorganisms or enzymes.

Generally, mechanical processes are done in order to change the particle size, shape and bulk density of the raw materials, such as the mechanical size reduction, which commonly is the first step in a biorefinery. On the other hand, for processes as adsorption and extraction, the aim is to separate and/or concentrate the substrates, intermediates or final products into their components [41]. The latter group includes a broad range of chemical reactions such as hydrolysis and transesterification processes [12, 41].

2.4. Colombian Lignocellulosic Wastes as Potential Feedstocks for Valuable Products

Because of its geographic location and its wide range of climates, Colombia is the third-ranked country in the world in terms of biodiversity, housing approximately the 10% of the world species, and the fourth-ranked in terms of hydrographic resources [44]. According to the Departamento Administrativo Nacional de Estadística (DANE), from 2000 to 2009, agriculture contributed an increase of 27.4% to the Colombian Gross Domestic Product (GDP). Furthermore, the World Trade Organization reported that the agroindustrial exports represented for Colombia approximately 30% of its total exports [45]. Figure 4 shows the use share of the land in Colombia. It is estimated that approximately 44.77% of the continental surface of the country (11.174.800 hectares) is destined to agricultural activities [46]. Although Colombia counts with approximately 21,5 million ha that could be dedicated to agriculture, just 4,9 million ha are actually used. In contrast, only 20 million ha are suitable for livestock, but almost 38.6 million ha are currently used for this activity [47].

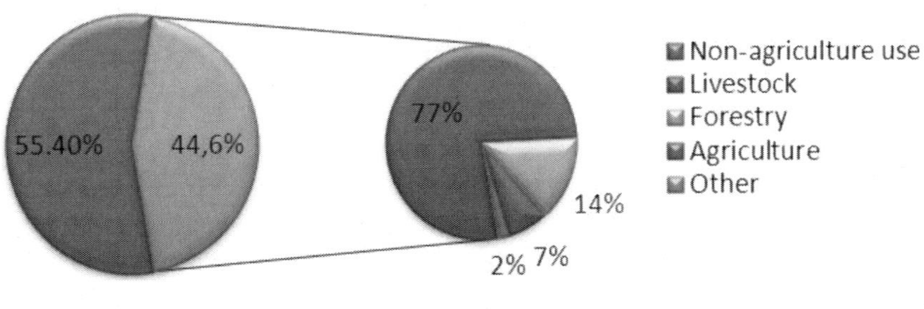

Figure 4. Land use share in Colombia [47, 48].

By 2005, 4.058.470 ha were dedicated to perennial and annual crops [46], which correspond to the 98% of the agriculture production while the remaining 2% corresponds to fallow areas [48]. In Colombia, the cutting and transformation processes of the different crops lead to the generation of more than 27 million tons of wastes per year [23, 46]. These agricultural crop residues can be classified as field and processing residues [46, 49]. Residues left in an agricultural field after harvesting the crop such as empty fruit bunches from palm plantations; discarded fruits, leaves and stems, straw and stalk and seed pods are considered field residues. On the other hand, materials obtained as byproducts after processing the crops such as husks, seeds, bagasse, peel and roots are classified as processing residues. According to the Ministerio de Minas y Energía [46], 75% of the total biomass produced by the agricultural sector in Colombia corresponds to crop residues, while processing residues represent the remaining 25%. Although 100% of the processing residues can be recovered to obtain valuable products such as energy and biofuels, a percentage of the field residues should be recycled onto the land. Field residues protect the soil from erosion, help to build organic matter and contribute the nutrients to get back to the soil. The percentage of residue removal depends on the specific local clima and soil conditions [50]. However, it is estimated that the residue removal can reach 35% without adverse effects on future yields [50].

3. Methodology Description

3.1. Orange Peel

Valencia OP was provided by Frugy S.A. (Manizales, Caldas, Colombia). OP was stored at -20°C before its analysis and processing within 5 days. To carry out its characterization, OP was air-dried during 24 hours (temperature of 18°C and a relativity humidity of 80%), and then oven-dried at 60°C during 48 hours to remove the excess of water [51]. Finally the dried OP was also milled up to 0.42 mm (Retsch model ZM 200, Retsch-Solutions in Milling & Sieving, Hann, Germany).

3.2. Characterization

OP characterization, performed in duplicates, comprised determination of extractives [52], moisture and total solids [53] and the content of insoluble solids [54], fixed and volatile solids [55]; cellulose, hemicellulose and lignin [56]; and, byproducts and inhibitors [57]. These characterization procedures were performed following the Standard Biomass Laboratory Procedures from the National Renewable Energy Laboratory (NREL). Besides, the elemental analysis and the nutrients composition were assessed.

3.2.1. Reagents

Analytical grade anhydrous xylose, anhydrous arabinose, calcium carbonate and furfural were purchased from Sigma-Aldrich (San Luis, MO); anhydrous glucose, anhydrous mannose and anhydrous galactose from Fluka (Buchs, Switzerland); anhydrous ethanol, sulfuric acid (96%), nitric acid (65%) and hydrogen peroxide from Panreac (Barcelona, Spain); formic and acetic acids from Riedel de Haën (Morristown, NJ); and, hydroxymethylfurfural (HMF) from Alfa Aesar (Ward Hill, MA).

3.2.2. Elemental Analysis

Elemental analysis of the dried samples (0.5 g) was performed using an Organic Elemental Analyzer Flash EA 1112 (Thermo Scientific, Waltham, MA) to obtain the total nitrogen, carbon, hydrogen, sulfur and oxygen contents.

Table 1. Conditions to carry out the digestion for nutrient determination

Stage	Max. Power (W)	% Power	Ramp (min)	Temperature (°C)	Hold (min)
1	1600	90	15	200	5
2	1600	90	1	210	5
3	1600	90	1	220	5

3.2.3. Nutrients

Potassium (K), phosphorous (P), sodium (Na), calcium (Ca) and magnesium (Mg) were determined by Inductively Coupled Plasma Mass Spectrometry (ICP-MS) 7500a (Agilent Technologies, Santa Clara, CA). Firstly, a dry basis sample was pretreated by microwave digestion using a CEM MARS Express Microwave Accelerated Reaction System (CEM Corporation, Matthews, NC), with adjustable power between 0-1600 watts and a frequency of 2455 MHz. The digestion breaks down the sample matrix leaving the nutrients in solution. The sample (0.5 g) was placed into a 55 ml Xpress tube (220°C and 500 psi maxima) with 7 ml of nitric acid (70%) and hydrogen peroxide (30%) and heated as described in Table 1. After the digestion, the resulting liquid was diluted in deionized water up to 100 ml using a volumetric flask and analyzed through ICP-MS.

3.3. Process Design Approach

Three stand-alone processes were simulated: (1) biogas and biofertilizers production through AD; (2) gibberellic acid (GA_3) production through solid state fermentation (SSF); and, (3) bioenergy production through gasification. Moreover, a biorefinery scheme was proposed, based on the OP experimental composition and previously published works [58], in order to compare the economic and environmental performance of the three processes proposed.

Yields for the alkaline pretreatment of the raw materials with KOH for AD and SSF were taken from Liu et al. [59]. The AD process was modeled using the kinetic model and yields reported by Dai et al. [60]. GA_3 production was modeled using the kinetics reported by Gelmi et al. [61]. Yields for limonene and pectin extraction processes were taken from Cerón and Cardona [21]. The processes simulated in this chapter are described in detail in the following sections.

3.4. Process Description

3.4.1. Anaerobic Digestion Process

The solid substrate was first dried to 8% moisture content and milled to a 40 mm particle size. Then, for the start-up of the AD process, the milled OP was mixed with the activated sludge from an anaerobic reactor for municipal solid waste to obtain 22% of total solid (TS) content. Then, urea was added to the solid substrate to obtain a C/N ratio of 30. The digestion process was modeled according to the kinetic model described by Dai et al. [60]. After 60 days the slurry obtained was separated into the solid (25% moisture content) and liquid phases (biosol and biol, respectively). Part of the liquid slurry (enough to obtain 22% of TS content) was recycled to the anaerobic digester to be used as inocula in the process. The remaining liquid phase was used as foliar fertilizers while the solid would be used as soil fertilizer. It was assumed that the biogas is directly used for cooking and then desulphurized to a maximum concentration of 10 ppm H_2S.

3.4.2. Solid State Fermentation Process

OP was firstly pretreated at 20°C for 24 h with potassium hydroxide (40% on dry weight basis of the feedstock) with a solid to water ratio of 1:10 w/v [59]. Then, the resulting stream was filtered (until 10% moisture) and the solids sent to SSF process and mixed with a mineral salt solution in a 1,6:1 liquid to solid ratio. After adjusting the initial moisture of the medium to 60% sterilization at 121°C for 15 min would be used before inoculation with *Fusarium solani* and subsequent fermentation at 30°C for 10 days [62]. GA_3 production was modeled using the kinetic expressions reported by Gelmi et al. [61]. Gibberellins were extracted from the SSF by adding water at a 1:20 initial solid to liquid ratio and shaking at 150 rpm for 2 hours at 30°C [63]. The slurry obtained was centrifuged, filtered and sent to a triple effect evaporator [64] for removing water. The resulting stream was subjected to a crystallization process at 20° C [64]. The cake obtained by centrifugation of the liquid stream was subsequently sent to a dryer obtaining the GA_3 powder with 0,1% moisture content while the supernatant was recycled to the process.

3.4.3. Energy Cogeneration

OP, from juice extraction process, with a moisture content of 85.9% was dried up to obtain a moisture of 15% [65] and then it was milled up to a particle size of 80 mm [65]. Then the dried material was gasified with air at an equivalent ratio of 0,25 [66] at 60 bar. After, the obtained syngas was depressurized up to 1 bar to obtain electricity.

3.4.4. Limonene Extraction

Limonene content in the essential oil from orange peel is approximately 98% [58]. Orange peel milled to reduce particle size was sent to a steam distillation at 90ºC to extract the essential oil and obtaining oil-water emulsion containing 95.6% essential oil by weight [58]. The steam distillation can reach yields higher than those obtained with the vacuum distillation process [58]. The resulting stream was subjected to a liquid-liquid separation stage at 25°C to separate the essential oil [58].

3.4.5. Pectin Extraction

Output solids from the oil extraction process were assumed be sent to a pectin extraction process by hydrolysis with citric acid during 60 minutes at 80°C, as described by Cerón and Cardona [58]. Pectin was then decanted using ethanol at 95% in a ratio 1:2 output stream to ethanol ratio [58].

3.5. Process Simulation

In order to obtain the mass and energy balances to perform the economic and environmental assessments, must be define, structure, specify and simulate the various technological schemes proposed to transform the OP into biofertilizers, bioenergy and other value-added products. The commercial package Aspen Plus v8.2 (Aspen Technology Inc.) was the main tool used. The most complex and detailed technological schemes were obtained through rigorous simulations, which involve sensitivity analysis and search of optimal operation conditions. Besides, the specialized package Matlab was used for performing mathematical calculations including those required for kinetic analysis. Thermodynamic properties for conventional molecules were obtained from [67] and the NIST Database. Meanwhile, the biomass was introduced into the simulator database according to the physicochemical compositions determined following the methodology described. The thermodynamic models used for the simulation were UNIFAC Dortmund method and Soave Redlich Kwong with the Bosto Mathias (RKS-BM) modification method [68, 69] for liquid and vapor phases, respectively. On the other hand, for simulations involving ethanol, the NRTL model was employed using specific physical-property obtained from [70, 71].

3.6. Process Assessments

3.6.1. Energy Consumption

Energy consumption estimations were performed using the mass and energy balances generated by Aspen simulator. Then, thermal energy required by heat exchangers and re-boilers and the electric energy needs of the pumps, compressors, mills and other equipment were calculated.

3.6.2. Economic Assessment

Due to the OP low availability in Colombia, the economic analysis of AD and gasification processes were performed following the one recommended by FAO [72] to evaluate the total production cost of biogas and biofertilizers and electricity, respectively. For the SSF process the commercial package Aspen Economic Analyzer V8.2 was implemented. Specific parameters regarding to Colombia conditions such as income tax, annual interest rate and labor salaries, among others, were incorporated in order to calculate the production costs per unit for the different products (Table 2). In Table 3 and Table 4 the market prices of the raw materials and products used in the economic evaluation are respectively summarized.

This analysis was estimated in US dollars for a 10-year period using a straight-line capital depreciation method. Besides, the net and unit profit margin of the proposed processes and its products, respectively, as well as the unit production cost in each case are shown as parameters for performing the economic assessment.

3.6.3. Environmental Assessment

The environmental assessment included the potential environmental impact (PEI), a relative measure of the potential for a chemical to have an adverse effect on human health and the environment and the greenhouse gases (GHG) emissions associated with the proposed processes.

Table 2. Economic parameters at Colombian conditions

Parameter	Unit	Value
Income tax	%	33
Annual interest rate	%	17
Operator labor cost	USD/h	2.14
Supervisor labor cost	USD/h	4.29
Electricity (interconnected zones)	USD/kWh	0.10
Electricity (non-interconnected zones)*	USD/kWh	0.20
Potable water	USD/m^3	1.25
Fuel	USD/MWh	26.81

* National average [73].

Table 3. Prices of the raw materials used for performing the economic evaluation

Raw material	Unit	Price	Reference
OP	USD/Ton	5.00	Local price
Urea	USD/Ton	369.80	[74]
KOH	USD/kg	0.88	[74]
Low P. steam (3 bar)	USD/Ton	1.57	[75]

Table 4. Prices of the products used for performing the economic evaluation

Products	Unit	Price	Reference
Liquid fertilizer (biol)	USD/Ton	630.00	[76]
Solid fertilizer (biosol)	USD/Ton	320.00	[76]
Gibberellic acid	USD/kg	100.00-500.00	[77]
Xylitol	USD/kg	7.95	[78]
Ethanol	USD/L	1.24	[79]
Furfural	USD/kg	2.00	[80, 81]
Orange Oil (Essential)	USD/kg	3.25	[74]
Pectin	USD/kg	12.62	[74]

Potential Environmental Impact (PEI)

The Waste Reduction Algorithm WAR, developed by the National Risk Management Research Laboratory of the U.S. Environmental Protection Agency (EPA) was used as the PEI calculation method. The PEI for a given mass or energy quantity is defined as the effect that energy and mass have on the environment if they are arbitrary discharged. The environmental impact is a quantity that cannot be directly measured, however, it can be calculated from measurable indicators. Environmental impact categories used in the WAR GUI software include: Human toxicity by ingestion (HTPI), human toxicity by dermal exposition or inhalation (HTPE), aquatic toxicity potential (ATP), Global warming (GWP), Ozone depletion potential (ODP), Photochemical oxidation potential (PCOP) and acidification Potential (AP). This tool considers the impact by mass effluents and the impact by energy requirements of a chemical process, based on the energy and mass balances generated in Aspen Plus. Then the weighted sum of all impacts ends into the final impact per kg of products.

Greenhouse Gas (GHG) Emissions

Another important environmental parameter is the GHG emissions associated with a particular process. This procedure was completed following the IPPC Guidelines [82]. GHG emissions were calculated using equivalent factors of 14 for CH_4, 4.5 for CO, 196 for NO_x. Also GHG emissions were calculated for energy needs with no integration of cogeneration. The external energy source assumed was charcoal with an emission factor of 94600 kg CO2-e/TJ [82]. Nevertheless, when the process configurations were coupled with cogenerations systems the GHG emissions correspond to the mass balance in the exhausted gases from gasification.

4. RESULTS AND DISCUSSION

4.1. OP Characterization

The physicochemical characterization of the OP was performed according to the experimental procedure above described. The physicochemical characterization of the OP in dry basis is presented in Table 5. This raw material has a moisture content of 4.46 ± 0.01%, which is according to the moisture reported by [83]. The structural sugars present in OP are glucose (12.73 ± 0.61%), xylose (6.17 ± 0.09%), galactose (3.92 ± 0.03%), arabinose (3.90 ± 0.02%) and manose (0.49 ± 0.04%). Acid soluble and acid insoluble lignin contents are 2.50 ± 0.01% and 6.19 ± 0.18%, respectively. Meanwhile, acid insoluble ash content is 0.08 ± 0.01%. Water-soluble extractives represent the 91.97% of the total extractives and are composed by glucose (8.68 ± 0.15%), xylose (0.50 ± 0.04%), Galactose (1.66 ± 0.20%), arabinose (3.37 ± 0.42%), manitol (0.86 ± 0.0003%), xylitol (0.47 ± 0.07%) and acetic acid (1.39 ± 0.13%).

According to the results, total solids (TS) and volatile solids (VS) contents, important operation parameters for AD process, are 89.50 and 88.91%, respectively. According to the results, extractives content represents the major component of OP, with low contents of cellulose and hemicellulose. These results are in agreement with those reported in the literature where lignin, cellulose, hemicellulose and ash range between 0.84-2.10%, 9.21-13.61%, 6.10-10.5%, and 1.50-5.51% [58, 83-86], respectively.

Table 6 summarizes the ultimate analysis of the OP characterized. The results show that carbon and oxygen are the main elemental compounds of this feedstock, representing approximately 93% of the raw material with oxygen representing more than the 50% of the total. According to these results, C/N ratio for CH is 43.64. Results obtained in this research are in agreement with those reported by [87, 88].

Phosphorous, potassium, calcium, sodium and magnesium concentrations were determined for OP and the results are summarized in Table 7. According to these results, in Colombia approximately 58, 277, 325, 3 and 54 tons of P, K, Ca, Na and Mg are lost with the OP, annually.

Table 5. Orange peel physicochemical characterization in dry basis

Feature	Composition (%)	Deviation
Cellulose	11.57	0.59
Hemicellulose	12.93	0.11
Lignin	8.69	0.19
Extractives	38.89	0.95
Ash	2.39	0.01
O-Acyl groups	0.59	0.01

Table 6. Elemental analysis for the OP

Element	Composition (%)	Deviation
Nitrogen	0.95	0.02
Carbon	41.30	0.04
Hydrogen	6.51	0.04
Oxygen	51.24	0.10

Table 7. Nutrients concentration for OP

Nutrient	Concentration (mg/kg OP)	Deviation
P	1400.11	53.63
K	6671.34	105.84
Ca	7833.94	149.48
Na	70.51	2.24
Mg	1303.11	37.44

These results have been used to simulate the stand-alone processes of biofertilizers, gibberellic acid and electricity production, as well as the biorefinery schemes presented in this chapter.

4.2. Techno-Economic Assessment

4.2.1. Gasification

The gasification process is presented as a base case to compare its economic behavior with the AD and SSF processes. Since the availability of this raw material is low in the country, low-scale gasification was considered for three different processing capacities: 500, 2000 and 5000 kg. The gasification process was simulated according to the procedure described in section 3.4.3. The commercial software Aspen Plus was used to generate the results presented in Table 8.

For OP gasification yields of 11 and 200 kW/Ton in wet and dry basis were obtained, respectively. The annualized cost of the gasification system and its share are summarized in Table 9 and Figure 5.

Table 8. Electricity production for gasification of OP

	OP processing capacity (kg)		
	500	1000	5000
Electricity (kWh)	5.93	23.73	59.34

The economic analysis for the low-scale gasification process was carried out based on the procedure described by FAO for rural gasification [65], considering 3600 working hours/year. For low-scale gasification, the production cost is always higher than the national average price for non-interconnected zones of 0.2 USD/kWh [73]. In this case, the economic parameter that affects the most the production cost is the raw materials and its effect increases while the processing capacity does.

Table 9. Annualized production cost of the low-scale gasification process of OP

Economic Parameter	OP processing capacity (kg/h)		
	500	1000	5000
	Thousand USD/year		
Capital depreciation	6.37	19.48	28.22
Raw materials	9.00	36.00	90.00
Operation	2.72	5.86	10.86
Total	18.09	61.34	129.09
Electricity (USD/kWh)	0.60	0.52	0.43
Profitability index	-0.52	-0.57	-0.92

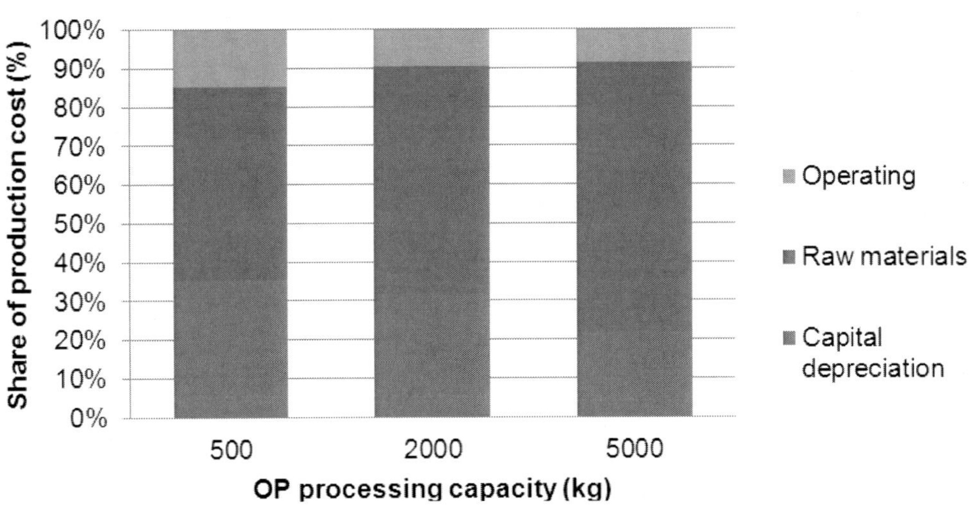

Figure 5. Share of production cost for the gasification process of OP.

According to UPME [89], the basic subsistence electricity consumption in Colombia per house ranges between 1560 to 2076 kWh-year. It means that for gasification processes, the yearly electricity demand of 5-7, 13-18, 20-30 houses can be supplied for 500, 1000 and 5000 kg of OP gasified, respectively, during 15 hours per day and 5 days a week. It was assumed that power losses due to operation represent the 15% and power losses in distribution network are 6% [65]. Although the electricity production cost under these conditions is significantly higher than the national average market price for non-interconnected zones in Colombia, it should be considered that in most of these zones the electric service is available just for 4-6 hours per day, while this study considers 15 working hours per day. Moreover in some non-interconnected zones of the country, such as Vaupés, the electric service can even cost 0.32 USD/kWh [73].

4.2.2. Anaerobic Digestion

For AD, also three production capacities: 500, 1000 and 5000 kg for biofertilizers production were considered. In the process liquid and solid biofertilizers and biogas were obtained. The process described in section 3.4.1 was simulated using the Aspen Plus commercial software to obtain the mass and energy balances summarized in Table 10.

The results indicate that the 95% of the OP to the AD process was recovered as biofertilizers. This is in accordance with the value of 85-90% reported by [90]. Besides, the liquid phase represented approximately the 83% of the obtained bioslurry, evidencing a VS recovery of 44% that corresponds to the VS recovery of 43.1% reported by [60]. Based on the nutrient balance of the AD process, the calculated nutrient composition of both liquid and solid biofertilizers is summarized in Table 11. Phosphorous and calcium presented a considerable increase due to its high content in the anaerobic sludge. Meanwhile, nitrogen content increases because of the use of urea to improve the C/N ratio during the AD. The nitrogen content of the solid biofertilizer was similar to that of ammonium sulfate (210 g/kg), and the double of di-ammonium and mono-ammonium phosphate (110 g/kg), while the phosphate content is approximately the 25% of di-ammonium and mono-ammonium phosphate (480 g P_2O_5/kg). Moreover, the calcium content is also approximately the same of the Ca content of calcium nitrate (210 g/kg).

Table 10. Mass and energy balances for the AD process of OP

	OP processing capacity (kg)		
	500	2000	5000
Other raw materials			
Urea (kg)	5.12	20.49	51.23
Activated sludge (kg)*	223.68	894.73	2236.84
Water (L)	79.00	327.00	789.00
Products			
Biol (Kg)**	239.01	956.02	2390.05
Biosol (kg)	47.54	190.18	475.44
Biogas (m^3)	1.50	5.96	15.03

* Starter culture, required as inocula for the start-up of the AD process.

** Net production, part of which is recycled as inocula for other processes.

After desulphurization, the biogas was composed by 85% of methane, 10% of CO_2, 1.6% of N_2, 2.4% of H_2 and 0.8% of water. For all the OP processing capacities, the biogas energy content ranges between 4.1-4.2 kWh/m^3. This energy content corresponds to almost a half litter of diesel (10 kwh/L), a half cubic meter of natural gas (8.8 kWh/m^3), a half kg of coal (8.1 kWh/kg) and almost 1 kg of dry wood (5.3 kWh/kg) [91]. Here it is considered that the biogas is directly used for cooking. According to [92] gas consumption for cooking per person and per meal ranges between 150 to 300 L of biogas. Then, the biogas production rates allow cooking 1-3, 3-7 and 6-12 meals for 65, 150 and 250 kg of OP digested, respectively.

Table 11. Nutrient balance for AD of OP

Nutrient	Nutrient content (g/kg)				
	Inlet			Outlet	
	OP	Sludge [93]	Urea	Biosol*	Biol*
N	9.33	40.7	460	239.76	0.28
P	0.44	25.4	-	131.69	0.50
K	0.37	1.00	-	61.32	2.69
Ca	0.63	25.30	-	199.46	0.39
Mg	0.68	4.30	-	31.87	0.41
Na	0.46	0.30	-	1.78	0.07

* Distribution of nutrients calculated according to the literature [94].

Table 12. Annualized production cost for AD of OP

Economic Parameter	OP processing capacity (kg)		
	65	150	250
	USD/year		
Capital depreciation	33.76	32.23	33.77
Raw materials	28.91	63.21	285.59
Operation	2.83	5.29	17.91
Total	65.50	120.59	392.24
Production cost (USD/Ton)			
Biol	481.78	221.85	288.67
Biosol	74.27	34.18	44.48
Profitability index	1.10	4.07	5.52
Payback period (years)	8.00	2.00	1.00

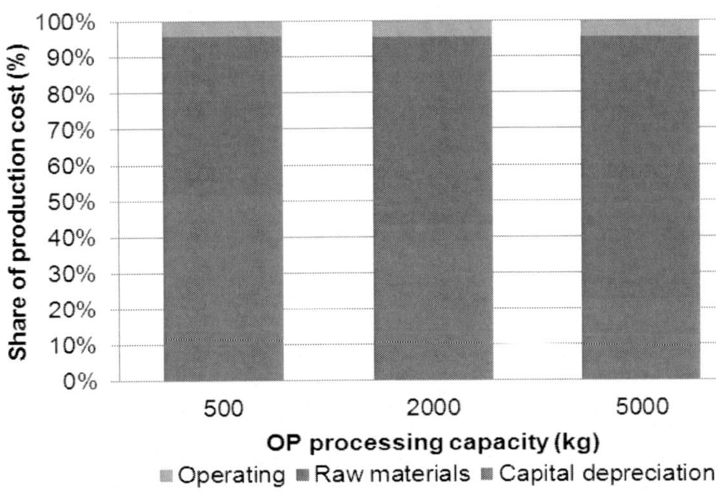

Figure 6. Share of production cost for low-scale biofertilizers production.

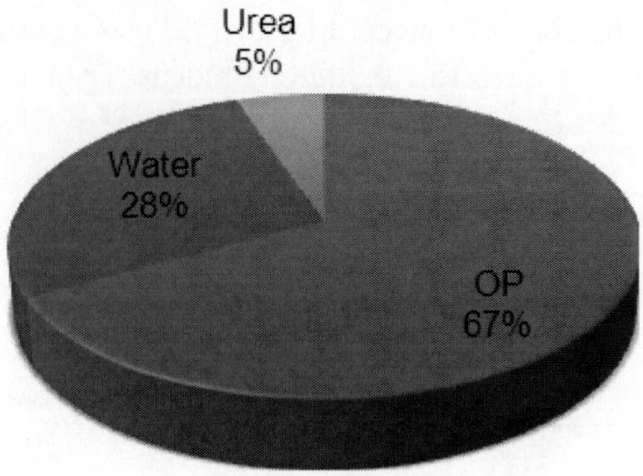

Figure 7. Distribution costs of the raw materials for AD of OP.

Usually AD in rural communities is carried out in tubular or polyethylene bag reactors that consist on digesters built from two layers of polyethylene plastic in a tubular form [72]. This reactor is relatively inexpensive and easy to construct and have a lifetime of about 10 years. The economic assessment of AD plants was carried out following the procedure described by FAO [72]. The annualized production costs per ton of biosol and biol are summarized in Table 12, while its share is presented in Figure 6.

For AD process, the economic factor that affects the most the productions cost is the capital depreciation, although while the OP processing capacity increases this effect is lower and the effect of the raw material increases as well. The distribution cost of the raw material is presented in Figure 7, where OP represents the 67% of the total costs. For 1000 and 5000 kg/h of processing capacity, the profitability index is above the unit, indicating economic feasibility of the process. This is in agreement with the NPV dependence presented in Figure 8. For both biol and biosol the production cost per ton is lower than the market prices of 630 and 320 USD/ton [76].

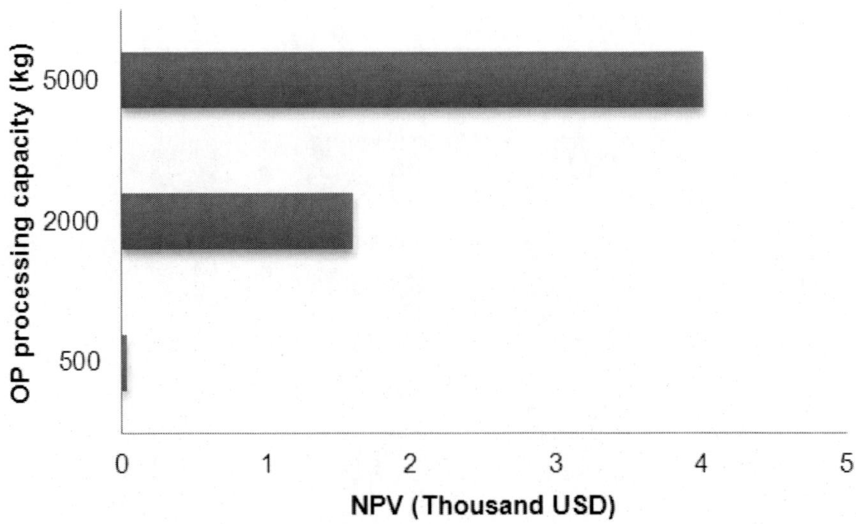

Figure 8. Influence of the OP processing capacity on the NPV for low-scale AD.

Table 13. Mass and energy balances for low-scale SSF of OP to produce GA_3

	OP processing capacity (kg)		
	500	1000	5000
Other raw materials			
KOH (kg)	0.88	3.52	8.81
Water for pretreatment (m^3)	0.35	1.41	3.52
Urea (kg)	0.45	1.79	4.47
Water for fermentation (m^3)	0.15	0.59	1.48
Water for separation (m^3)	1.41	5.64	14.10
Product			
GA_3 (g)	49.00	98.00	196.00
Energy requirements			
Electricity (kWh)	62.86	65.45	69.83
Heating (MWh)	1.22	2.44	4.89
Cooling (MWh)	0.04	0.26	0.52

4.2.3. Solid State Fermentation

In this chapter, the goal of the SSF is the production of gibberellic acid (GA_3). Also three production capacities were evaluated: 500, 1000 and 5000 kg. The process was simulated based on the procedure described in section 4.3.2. The mass and energy balances for this process are summarized in Table 13.

The results indicate that the productivity of gibberellic acid from OP is 0.45 and 1.35 g of GA_3 per kg of wet and dried OP, respectively. These yields are consisted with the reported by [95] of 1.22 g of GA_3 per kg of dried substrate. Production costs and share associated to this process are presented in Table 14 and Figure 9, respectively.

Table 14. Production costs of SSF of OP to produce GA_3

Economic Parameter	OP processing capacity (kg)		
	65	150	250
	Thousand USD/year		
Capital depreciation	659.57	668.06	689.66
Raw materials	46.76	147.03	467.57
Utilities	96.46	98.54	102.04
Operation	267.47	278.09	308.06
Total	1070.26	1191.71	1567.34
GA_3 production cost (USD/kg)	3185.30	1520.04	999.58
Selling price: 100 USD/kg			
Profitability index	0.02	0.04	0.07
Payback period	-	-	-
Selling price: 500 USD/kg			
Profitability index	0.10	0.22	0.36
Payback period	-	-	-

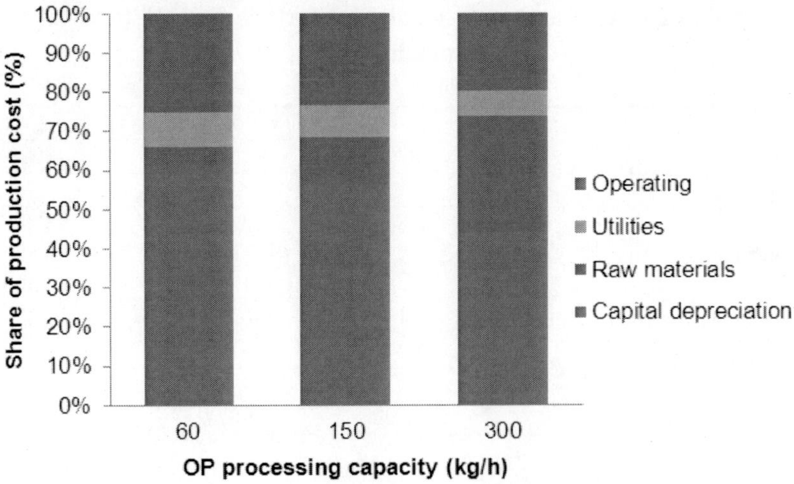

Figure 9. Share of production cost for GA_3 production from OP at low-scale.

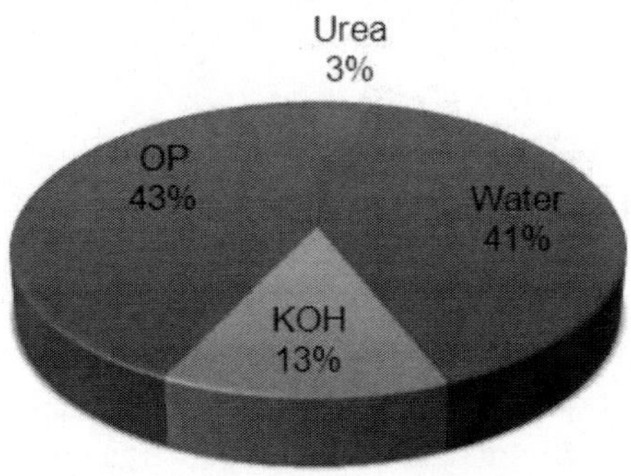

Figure 10. Distribution costs of the raw materials for SSF of OP.

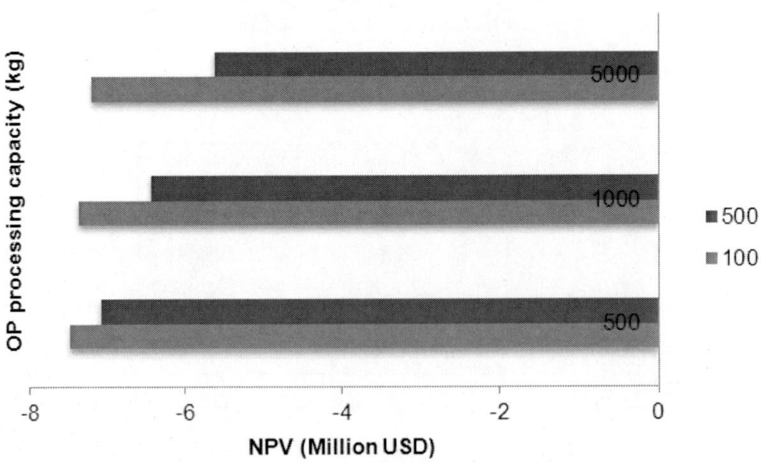

Figure 11. Influence of the OP processing capacity on the NPV for low-scale SSF.

For SSF process, the economic factor that affects the most the production cost is the capital depreciation. The distribution cost of the raw material is presented in Figure 10. For GA_3 production process, an additional sensitivity analysis regarding to the market price was developed. This analysis was considered based on the large range of prices found for this product (100-500 USD/kg). The profitability index and the payback period for SSF show that the process is not economic feasible at low scale since for all the cases profitability index is lesser than 1 and the payback period is into the first 10 years. This is in accordance with the NPV dependence presented in Figure 11.

4.2.4. Comparative Analysis between Technologies

As above mentioned, the base case for the economic comparison is the electricity production through gasification. The comparative analysis has been carried out based on the profitable index. For low-scale processes, the economic comparison is showed in Figure 12.

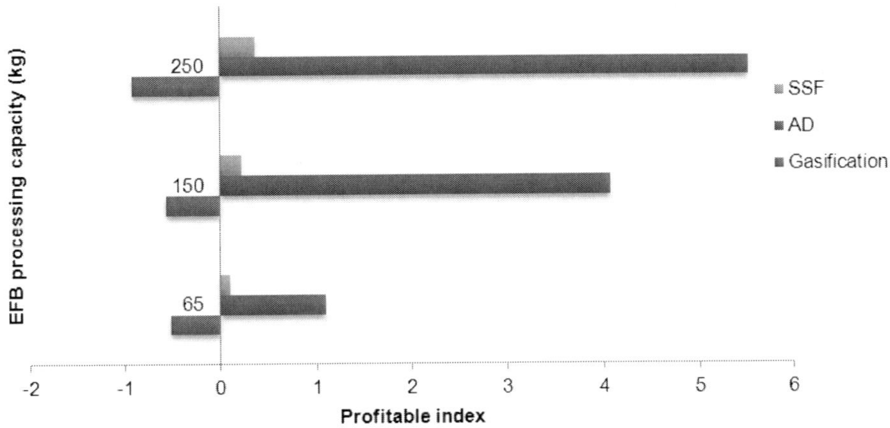

Figure 12. Economic Comparison between technologies for OP processing.

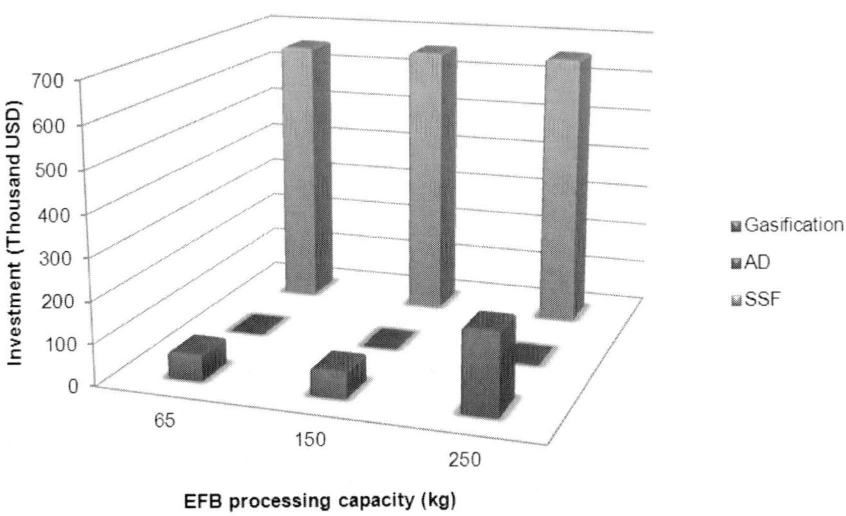

Figure 13. Investments cost comparison between technologies for OP processing.

According to Figure 12, for OP processing only the AD process is economically feasible for the processing capacities here considered, since for gasification and SSF the profitability indexes are negative and lower than the unit, respectively. The AD process presents the highest profitable indexes due to the main products here considered are produced in large quantities, instead of the GA_3 presents a very low yield and then the production volume is very low. Moreover, technology for producing GA_3 is more specialized and, hence more expensive. On the other hand, the low values for the profitability index in the gasification process are mainly due to the high investment cost (See Figure 13).

4.2.5. Environmental Assessment

The environmental assessment has been carried out as the comparison between processes, also considering the gasification process as the base case. For this evaluation, the WAR software was used to calculate the Potential Environmental Impact (PEI) per unit of product. Besides, the greenhouse gas (GHG) emissions have been also compared for the three technologies.

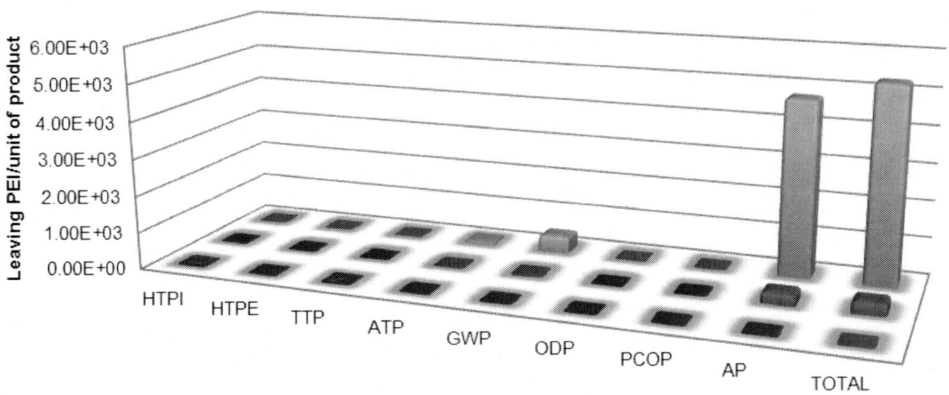

	HTPI	HTPE	TTP	ATP	GWP	ODP	PCOP	AP	TOTAL
Gasification	4.07E+00	5.09E-02	4.07E+00	2.87E-06	4.02E-03	0.00E+00	5.29E-02	0.00E+00	8.23E+00
AD	1.48E-02	4.04E-02	1.48E-02	5.12E+00	2.97E+01	1.13E-04	2.18E-02	3.08E+02	3.43E+02
SSF	2.29E-01	6.24E-01	2.29E-01	7.95E+01	4.60E+02	1.74E-03	3.36E-01	4.76E+03	5.30E+03

(a) Leaving PEI/unit of product

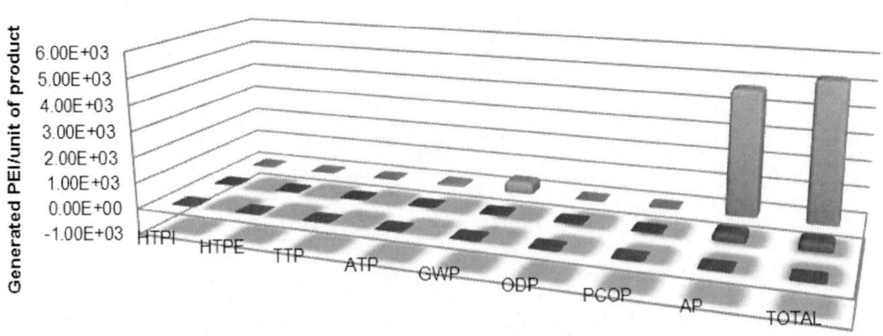

	HTPI	HTPE	TTP	ATP	GWP	ODP	PCOP	AP	TOTAL
Gasification	3.82E+00	2.55E-03	3.82E+00	-1.99E-03	4.02E-03	0.00E+00	-2.83E-02	0.00E+00	7.60E+00
AD	-5.77E-03	3.63E-02	-5.77E-03	5.12E+00	2.97E+01	1.13E-04	1.64E-02	3.08E+02	3.43E+02
SSF	2.27E-01	6.24E-01	2.27E-01	7.95E+01	4.60E+02	1.74E-03	3.35E-01	4.76E+03	5.30E+03

(b) Generated PEI/unit of product

Figure 14. Comparison of PEI per unit of product between technologies for low-scale OP processing.

4.2.6. Potential Environmental Impact

For the gasification process, the PEI and the GHG are calculated based on the electricity produced (PEI/kWh). Meanwhile for AD and SSF the units of product are the biofertilizers and the GA_3, respectively. The PEI per unit of product for the processing of OP for the three processes is presented in Figure 14.

Between the three processes, the SSF presents the higher both leaving and generated potential environmental impact. This behavior can be explained because of the large quantity of residues generated during the process that are no longer treated, such as waste water with traces of sugars and lignin-rich solid residues, affecting considerably the Acidification Potential (AP). Moreover the energy required during the SSF is higher than the requirements for AD and gasification. On the other hand, during gasification just the hot exhausted gases are considered residues, while during the AD process no residues are generated.

4.2.7. Green House Gas Emissions

Additionally, the GHG emissions were also calculated and compared between the three processes. The GHG emissions were calculated per kWh of produced electricity, ton of produced biofertilizers and kg of gibberellic acid, for gasification, AD and SSF, respectively. The results of the GHG assessment for low and high-scale OP processing are presented in Figure 15.

According to Figure 15, the process that presents the higher GHG emissions is the gasification of OP, followed by the anaerobic digestion. This can be explained by the release of gases to the environment during these two processes. Even though during the SSF process CO_2 and O_2 are released to the environment, there are very low quantities. The real environmental concerns related to the SSF process are the large quantities of liquid and solid residues that are produced.

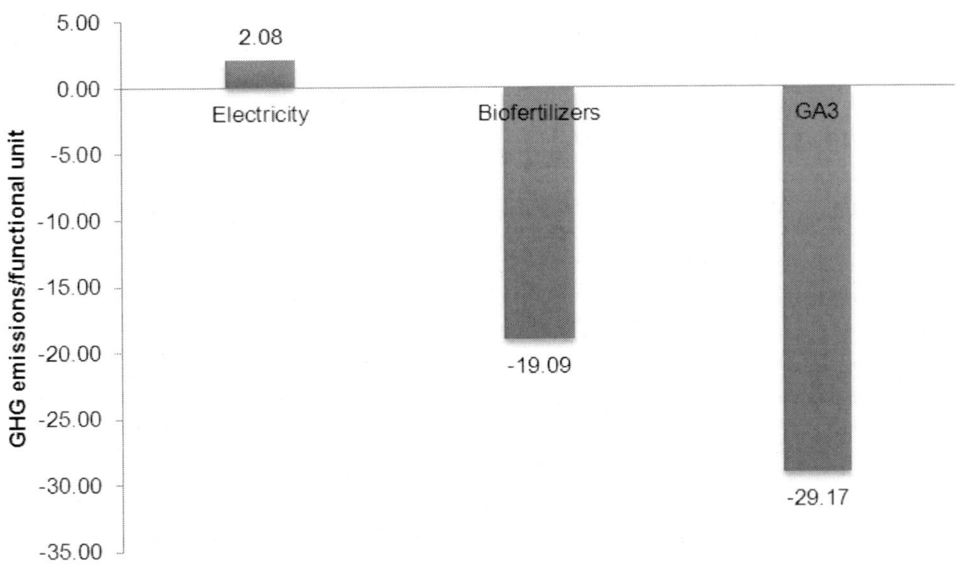

Figure 15. GHG emissions for low-scale processing of OP.

4.2.8. Biofertilizers and Gibberellic Acid Production under the Biorefinery Concept

According to the characterization here presented and the one reported by [21], the first scenario, limonene, pectin, biofertilizers and bioenergy are considered. The second scenario considers the production of limonene, pectin and gibberellic acid. For simulation purposes, 5 Ton/h of wet OP has been considered as raw material.

In the first step, orange peel is submitted to a milling process in order to reduce it size. Then it is sent to a steam distillation at 90ºC to extract the oil. Then, the Output solids from the oil extraction process are sent to a pectin extraction process. For the first scenario the solid obtained from the pectin extraction process, is sent to the anaerobic digestion process, to obtain biol and biosol. Moreover, the produced biogas is used for cooking in farmers. For the second scenario, the remaining solid is sent to the SSF process to produce gibberellic acid. The mass and energy balances of the described scenarios are summarized in Table 15. The annualized costs of the biorefinery, as well as the distribution cost are presented in Table 16.

Table 15. Mass and energy balances of the two proposed scenarios

Raw materials			Products		
OP	5.00	Ton	Limonene	3.72	kg
Ethanol	715.00	L	Pectin	27.92	kg
Citric acid	105.00	kg			
Water	2.07	m^3			
Scenario I			Scenario II		
Other raw materials					
Urea	51.22	kg	Water	8.35	m^3
Sludge	1.29	Ton	Urea	0.03	Ton
Water	0.57	m^3			
Products					
Biol	1.38	Ton	GA_3	5.70	kg
Biosol	0.28	kg			
Biogas	9.45	m^3			
Energy requirements					
Electricity	59.07	kWh	Electricity	67.91	kWh
Heating	2.23	MWh	Heating	12.35	MWh
Cooling	1.15	MWh	Cooling	0.41	MWh

Table 16. Annualized costs of the two proposed biorefinery scenarios

Economic parameter	Scenario I		Scenario II	
	Annualized costs (Million USD/year)	Share (%)	Annualized costs (Million USD/year)	Share (%)
Capital depreciation	0.41	29	0.55	32
Raw materials	0.66	46	0.75	43
Utilities	0.06	4	0.08	5
Operation	0.29	20	0.36	20
Total	1.43		1.73	
Profitability index	1.41		2.53	
Payback period	3.00		1.35	

Table 17. Unit production cost for both scenarios

Scenario I			Scenario II		
Limonene	0.09	USD/kg	Limonene	0.12	USD/kg
Pectin	0.15	USD/kg	Pectin	0.30	USD/kg
Biol	117.92	USD/Ton	GA_3	9.98	USD/kg
Biosol	4.25	USD/Ton			

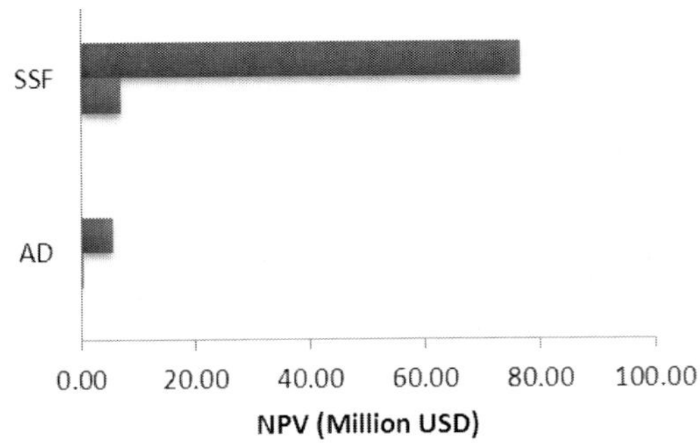

Figure 16. NPV for both scenarios.

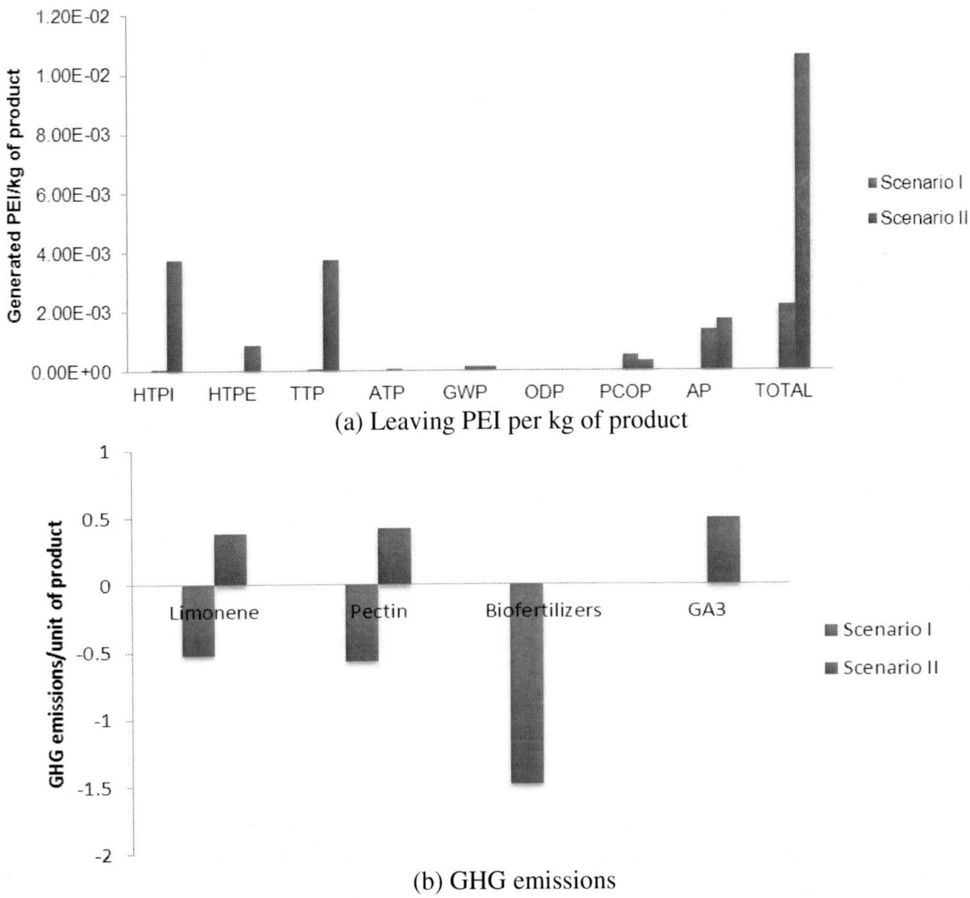

Figure 17. Environmental comparison of the scenarios.

For both scenarios the economic parameter that affects the most the production cost is the raw materials, mainly due to the ethanol and citric acid requirements for extracting the pectin. Production costs for each product in both scenarios are presented in Table 17.

For both scenarios, the profitability index is above the unit, indicating the economic feasibility of the biorefineries, which is confirmed by the NPV (Figure 16) and the advantage of ng small scale biorefineries to produce biofertilizers and GA_3. Moreover, the unit production costs for all the considered products are higher than the market prices, due to the low productivity of both scenarios, which is related with the low availability of the OP and its high moisture content.

The environmental comparison has been carried out based on the PEI and GHG emissions per kg of product. The mass allocation method has been used to develop the assessments. Results for the environmental comparison are presented in Figure 17. The leaving PEI is higher for scenario II, where GA_3 production through SSF was considered. It can be explained by the large quantity of liquid residues obtained during the SSF process. This result is confirmed by the GHG emissions, presenting negative GHG values the scenario I.

CONCLUSION

OP is a low-available raw material in Colombia and currently its processing to obtain valuable-products is not an established practice in the country. Despite of the logistic problems associated to its transport, because of its high moisture content and that is not concentrated in one region in the country, its utilization could represent a good economic and environmental alternative.

For low-scale processing, the best stand-alone process from the economic point of view is the biofertilizers production through anaerobic digestion, because of the low investment cost of this process. In the case of gasification, no any processing scale gives a good economic outlook, although it could be considered an alternative to solve out the energetic problems in non-interconnected zones where the electric service is reduced or even inexistent.

From the environmental point of view, at low scale processing of OP, the GA_3 process presents the worst performance because of the high quantity of water used during the process and that is considered as waste.

When the biofertilizers and GA_3 production process are considered into a biorefinery context together with other valuable products, the NPV value increases considerably, justifying this type of analysis.

REFERENCES

[1] DANE, Encuesta Nacional Agrícola 2009. 2011.

[2] Ministerio de Agricultura y Desarrollo Rural, Resultados evaluaciones agropecuarias municipales 2011, in Anuario estadístico de frutas y hortalizas 2007-2011 y sus calendarios de siembras y cosechas, Dirección de Política Sectorial and Grupo Sistemas de Información, Editors. 2012.

[3] J.C., R.R.A., Valencia; J.L., Restrepo; A., Uribe; J.E., Murillo, Cítricos: Cultivo, Poscosecha e industrialización, 2012, Corporación Universitaria Lasallista.

[4] Nassar, A.G., A.A. AbdEl-Hamied, and E.A. El-Naggar, Effect of Citrus by-Products Flour Incorporation on Chemical, Rheological and Organolepic Characteristics of Biscuits. *World Journal of Agricultural Sciences,* 2008. 4(5): p. 612-616.

[5] Cardona, C.A. and J.-S. Lee, Renewable Fuels. Developments in Bioethanol and Biodiesel Production. Universidad Nacional de Colombia, *Sede Manizales.* COLCIENCIAS., 2008.

[6] Pourbafrani, M., et al., Production of biofuels, limonene and pectin from citrus wastes. *Bioresource Technology*, 2010. 101(11): p. 4246-4250.

[7] Piarpuzán, D., J.A. Quintero, and C.A. Cardona, Empty fruit bunches from oil palm as a potential raw material for fuel ethanol production. *Biomass and Bioenergy*, 2011. 35(3): p. 1130-1137.

[8] Quintero, J.A., J. Moncada, and C.A. Cardona, Techno-economic analysis of bioethanol production from lignocellulosic residues in Colombia: A process simulation approach. *Bioresource Technology*, 2013. 139(0): p. 300-307.

[9] Tsai, S.-H., C.-P. Liu, and S.-S. Yang, Microbial conversion of food wastes for biofertilizer production with thermophilic lipolytic microbes. *Renewable Energy*, 2007. 32(6): p. 904-915.

[10] Gousterova, A., et al., Development of a biotechnological procedure for treatment of animal wastes to obtain inexpensive biofertilizer. *World Journal of Microbiology and Biotechnology*, 2008. 24(11): p. 2647-2652.

[11] Hernández, V., et al., Techno-economic and environmental assessment of an olive stone based biorefinery. *Resources, Conservation and Recycling,* 2014. 92(0): p. 145-150.

[12] Moncada, J., L.G. Matallana, and C.A. Cardona, Selection of Process Pathways for Biorefinery Design Using Optimization Tools: A Colombian Case for Conversion of Sugarcane Bagasse to Ethanol, Poly-3-hydroxybutyrate (PHB), and Energy. *Industrial and Engineering Chemistry Research*, 2013. 52: p. 4132 - 4145.

[13] Mussatto, S.I., et al., Techno-economic analysis for brewer's spent grains use on a biorefinery concept: *The Brazilian case. Bioresource Technology,* 2013. 148(0): p. 302-310.

[14] Rincón, L.E., V. Hernández, and C.A. Cardona, Analysis of technological schemes for the efficient production of added-value products from Colombian oleochemical feedstocks. *Process Biochemistry,* 2014. 49(3): p. 474-489.

[15] Rincón, L.E., et al., Optimization of the Colombian biodiesel supply chain from oil palm crop based on techno-economical and environmental criteria. *Energy Economics*, 2015. 47(0): p. 154-167.

[16] Naranjo, J.M., C.A. Cardona, and J.C. Higuita, Use of residual banana for polyhydroxybutyrate (PHB) production: Case of study in an integrated biorefinery. *Waste Management*, 2014. 34(12): p. 2634-2640.

[17] Naranjo, J.M., et al., Valorization of glycerol through the production of biopolymers: The PHB case using Bacillus megaterium. *Bioresource Technology,* 2013. 133(0): p. 38-44.

[18] Cerón, I.X., V. Hernández, and C.A. Cardona, Biorefinery Design for Value Added Products from Amazon Fruits, in 2nd Iberoamerican Congress on Biorefineries2013: Jaen, Spain

[19] Cerón, I.X., J.C. Higuita, and C.A. Cardona, Design and analysis of antioxidant compounds from Andes Berry fruits (Rubus glaucus Benth) using an enhanced-fluidity liquid extraction process with CO2 and ethanol. *The Journal of Supercritical Fluids*, 2012. 62(0): p. 96-101.

[20] Cerón, I.X., et al., Process synthesis for antioxidant polyphenolic compounds production from Matisia cordata Bonpl. (zapote) pulp. *Journal of Food Engineering*, 2014. 134(0): p. 5-15.

[21] Cerón, I.X. and C.A. Cardona, Evaluación del proceso integral para la obtención de aceite esencial y pectina a partir de cáscara de naranja. *Ingeniería y ciencia*, 2011. 7(13): p. 65-86.

[22] Duque, S.H. and C.A. Cardona, Plantain and banana fruit as raw material for glucose production. *Journal of Biotechnology*, 2014. 185, Supplement(0): p. S34.

[23] Duque, S.H., C.A. Cardona, and J. Moncada, Techno-Economic and Environmental Analysis of Ethanol Production from 10 Agroindustrial Residues in Colombia. *Energy & Fuels*, 2015. 29(2): p. 775-783.

[24] García, R., et al., Characterization of Spanish biomass wastes for energy use. *Bioresource Technology*, 2012. 103(1): p. 249-258.

[25] Boluda-Aguilar, M., et al., Mandarin peel wastes pretreatment with steam explosion for bioethanol production. *Bioresource Technology,* 2010. 101(10): p. 3506-3513.

[26] Carvalheiro, F., L.C. Duarte, and F.M. Gírio, Hemicellulose biorefineries: a review on biomass pretreatments. 2008.

[27] Grohmann, K., R. Cameron, and B. Buslig, Fractionation and pretreatment of orange peel by dilute acid hydrolysis. *Bioresource Technology,* 1995. 54(2): p. 129-141.

[28] Zhang, Y.H.P., Reviving the carbohydrate economy via multi-product lignocellulose biorefineries. *Journal of industrial microbiology & biotechnology,* 2008. 35(5): p. 367-375.

[29] Jørgensen, H., J.B. Kristensen, and C. Felby, Enzymatic conversion of lignocellulose into fermentable sugars: challenges and opportunities. *Biofuels, Bioproducts and Biorefining*, 2007. 1(2): p. 119-134.

[30] Octave, S. and D. Thomas, Biorefinery: Toward an industrial metabolism. *Biochimie,* 2009. 91(6): p. 659-664.

[31] Clark, J.H., R. Luque, and A.S. Matharu, Green Chemistry, Biofuels, and Biorefinery. *Annual Review of Chemical and Biomolecular Engineering,* 2012. 3: p. 183-207.

[32] Bozell, J.J., Feedstocks for the Future - Biorefinery Production of Chemicals from Renewable Carbon. *Clean,* 2008. 36(8): p. 641-647.

[33] Ng, R.T.L., D.H.S. Tay, and D.K.S. Ng, Simultaneous Process Synthesis, Heat and Power Integration in a Sustainable Integrated Biorefinery. *Energy & Fuels*, 2012. 26(12): p. 7316-7330.

[34] de Jong, W. and G. Marcotullio, Overview of Biorefineries based on Co-Production of Furfural, Existing Concepts and Novel Developments. *International Journal of Chemical Reactor Engineering*, 2010. 8(1).

[35] Liu, S., L.P. Abrahamson, and G.M. Scott, Biorefinery: Ensuring biomass as a sustainable renewable source of chemicals, materials, and energy. *Biomass and Bioenergy*, 2012. 39: p. 1-4.

[36] Fatih Demirbas, M., Biorefineries for biofuel upgrading: a critical review. *Applied Energy*, 2009. 86: p. S151-S161.

[37] Demirbas, A., Biorefinery technologies for biomass upgrading. Energy Sources, *Part A: Recovery, Utilization, and Environmental Effects,* 2010. 32(16): p. 1547-1558.

[38] Pham, V. and M. El-Halwagi, Process synthesis and optimization of biorefinery configurations. *AIChE Journal*, 2011. 58(4): p. 1212-1221.

[39] Cherubini, F. and S. Ulgiati, Crop residues as raw materials for biorefinery systems–A LCA case study. *Applied Energy*, 2010. 87(1): p. 47-57.

[40] Hughes, S.R., et al., Sustainable Multipurpose Biorefineries for Third-Generation Biofuels and Value-Added Co-Products, Biofuels - Economy, Environment and Sustainability, Z. Fang, Editor 2013.

[41] Hackl, R. and S. Harvey, Opportunities for Process Integrated Biorefinery Concepts in the Chemical Cluster in Stenungsund, 2010, Department of Energy and Environment. Division of Heat and Power Technology. Chalmers University of Technology: Göteborg, Sweden.

[42] Cherubini, F., The biorefinery concept: Using biomass instead of oil for producing energy and chemicals. Energy Conversion and Management, 2010. 51(7): p. 1412-1421.

[43] Aparcana, S. and A. Jansen, Estudio sobre el Valor Fertilizante de los Productos del Proceso "Fermentación Anaeróbica" para Producción de Biogás, G.P.P.E.a.E. Consultancy, Editor 2008.

[44] Mongabay, Colombia. Tropical Rainforest. Consulted August 2011, 2010. Avaliable from: http://rainforests.mongabay.com/20colombia.htm.

[45] Uribe, C.P., et al., Sembrando Innovación para la Competitivada del Sector Agropecuario Colombiano, 2011, Ministerio de Agricultura y Desarrollo Rural.

[46] Escalante, H., et al., Atlas del Potencial Energético de la Biomasa Residual en Colombia, 2007, Ministerio de Minas y Energía,.

[47] Restrepo, J.C., Una Política Integral de Tierras para Colombia, 2010, Ministerio de Agricultura y Desarrollo Rural.

[48] Ministerio de Agricultura y Desarrollo Rural and Corporación Colombia Internacional, Oferta Agropecuaria ENA - Cifras 2009. 2009.

[49] Hadar, Y., Sources for Lignocellulosic Raw Materials for the Production of Ethanol, in Lignocellulose Conversion, V. Faraco, Editor 2013, Springer Berlin Heidelberg. p. 21-38.

[50] Asakereh, A., et al., Spatial Analysis the Potential for Energy Generation from Crop Residues in Shodirwan, Iran. International Journal of u- and e- Service, *Science and Technology*, 2014. 7(1): p. 275-284.

[51] National Renewable Energy Laboratory, Determination of Sugars, Byproducts, and Degradation Products in Liquid Fraction Process Samples, in NREL Laboratory Analytical Procedures for standard biomass analysis2008, U.S. Department of Energy Office of Energy Efficiency and Renewable Energy.

[52] National Renewable Energy Laboratory, Preparation of Samples for Compositional Analysis, in NREL laboratory Analytical Procedures for standard biomass analysis2008, U.S. Department of Energy Office of Energy Efficiency and Renewable Energy.

[53] National Renewable Energy Laboratory, Determination of Extractives in Biomass. Laboratory Analytical Procedure., in NREL Laboratory Analytical Procedures for

standard biomass analysis2008, U.S. Department of Energy Office of Energy Efficiency and Renewable Energy.

[54] National Renewable Energy Laboratory, Determination of Total Solids in Biomass and Total Disolved Solids in Liquid Process Samples, in NREL Laboratory Analytical Procedures for standard biomass analysis2008, U.S. Department of Energy Office of Energy Efficiency and Renewable Energy.

[55] National Renewable Energy Laboratory, Determination of Insoluble Solids in Pretreated Biomass Material, in NREL Laboratory Analytical Procedures for standard biomass analysis2008, U.S. Department of Energy Office of Energy Efficiency and Renewable Energy.

[56] National Renewable Energy Laboratory, Determination of Ash in Biomass, in NREL Laboratory Analytical Procedures for standard biomass analysis2008, U.S. Department of Energy Office of Energy Efficiency and Renewable Energy.

[57] National Renewable Energy Laboratory, Determination of Structural Carbohydrates and Lignin in Biomass, in NREL Laboratory Analytical Procedures for standard biomass analysis2008, U.S. Department of Energy Office of Energy Efficiency and Renewable Energy.

[58] Cerón, I.X. and C.A. Cardona, Evaluación del proceso integral para la obtención de aceite esencial y pecitna a partir de cáscara de naranja. *Ingeniería y ciencia*, 2011. 7(13): p. 65-86.

[59] Liu, X., et al., Pretreatment of wheat straw with potassium hydroxide for increasing enzymatic and microbial degradability. *Bioresource Technology*, 2015. 185(0): p. 150-157.

[60] Dai, X., et al., High-solids anaerobic co-digestion of sewage sludge and food waste in comparison with mono digestions: Stability and performance. *Waste Management*, 2013. 33(2): p. 308-316.

[61] Gelmi, C., R. Pérez-Correa, and E. Agosin, Modelling Gibberella fujikuroi growth and GA3 production in solid-state fermentation. *Process Biochemistry*, 2002. 37(9): p. 1033-1040.

[62] Machado, C.M., et al., Gibberellic acid production by solid-state fermentation in coffee husk. *Applied Biochemistry and Biotechnology*, 2002. 102-103(1-6): p. 179-191.

[63] Rangaswamy, V., Improved Production of Gibberellic Acid by Fusarium moniliforme. *Journal of Microbiology Research*, 2012. 2(3): p. 51-55.

[64] Pfeifer, V.F., C. Vojnovic, and E.N. Heger, Itaconic Acid by Fermentation with Aspergillus Terreus. *Industrial & Engineering Chemistry Research*, 1952. 44(12).

[65] Food and Agriculture Organization of the United Nations, Gasification, in Bioenergy and Food Security Rapid Appraisal (BEFS RA)2014, Food and Agriculture Organization of the United Nations,.

[66] Chen, H., Lignocellulose Biorefinery Engineering: Principles and Applications2015: Elsevier Science.

[67] Wooley, R. and V. Putsche, Development of an ASPEN PLUS Physical Property Database for Biofuels Components, in Report No. NREL/MP-425-2068519961996, National Renewable Energy Laboratory: Golden, CO, USA.

[68] Batista, E., et al., Liquid−Liquid Equilibrium for Systems of Canola Oil, Oleic Acid, and Short-Chain Alcohols. *Journal of Chemical & Engineering Data*, 1999. 44(6): p. 1360-1364.

[69] Mathias, P.M., J.F. Boston, and S. Watanasiri, Effective utilization of equations of state for thermodynamic properties in process simulation. *AIChE Journal*, 1984. 30(2): p. 182-186.

[70] Wooley, R. and V. Putsche, Report NREL/MP-425-20685, 1996, National Renewable Energy Laboratory: Golden, CO, USA. p. 38.

[71] Wooley, R., et al., Lignocellulosic biomass to ethanol process design and economics utilizing co-current dilute acid prehydrolysis and enzymatic hydrolysis. Current and futuristic scenarios, in Technical Report NREL/TP-580-261571999, National Renewable Energy Laboratory: Golden, CO, USA. p. 123.

[72] Food and Agriculture Organization of the United Nations, Biogas - Small-Scale, in Bioenergy and Food Security Rapid Appraisal (BEFS RA)2014, Food and Agriculture Organization of the United Nations,.

[73] Flórez, J.H., D. Tobón, and G.A. Castillo, ¿Ha sido efectiva la promoción de soluciones energéticas en las zonas no interconectadas (ZNI) en Colombia?: Un análisis de la estructura institucional. *Cuadernos de administración*, 2009. 22(38): p. 219-245.

[74] ICIS, Indicative Chemical Prices A-Z. Available in: http://www.icis.com/chemicals/channel-info-chemicals-a-z/, 2012.

[75] Moncada, J., M.M. El-Halwagi, and C.A. Cardona, Techno-economic analysis for a sugarcane biorefinery: Colombian case. Bioresource Technology, 2012. In Press.

[76] Bioabonos Morines Alto. Venta de biol y biosol. 2015 [cited 2015 March, 20]; Available from: http://bioabonos.blogspot.com/p/venta-de-biol-y-biosol.html.

[77] Alibaba. Gibberellic acid. 2015 [cited 2015 January 12]; Available from: http://www.alibaba.com/showroom/gibberellic-acid.html.

[78] xylitolcanada, Xylitol price list. Available in: http://www. xylitolcanada.com, 2012.

[79] Fedebiocombustibles, Etanol anhidro de caña. Cifras Informativas del Sector Biocombustibles, 2011.

[80] Fuente-Hernández, A., et al., Biofuels and Co-Products Out of Hemicelluloses, in Liquid, Gaseous and Solid Biofuels - Conversion Techniques, Z. Fang, Editor 2013.

[81] Xiang, Z. and T. Runge, Co-production of feed and furfural from dried distillers' grains to improve corn ethanol profitability. *Industrial Crops and Products*, 2014. 55(0): p. 207-216.

[82] Eggleston, H., 2006 IPCC Guidelines for National Greenhouse Gas Inventories. *Forestry,* 2006. 5(OVERVIEW): p. 1-12.

[83] Adewole, E., et al., Phytochemical Constituents and Proximate Analysis of Orange Peel (citrus Fruit). *Journal of Advanced Botany and Zoology,* 2014. 1(3): p. 1-2.

[84] Miranda, R., et al., Characterization of Pyrolysis Products Obtained During the Preparation of Bio-Oil and Activated Carbon, in Lignocellulosic Precursors Used in the Synthesis of Activated Carbon - Characterization Techniques and Applications in the Wastewater Treatment, V. Hernández, Editor 2012, InTech.

[85] Ververis, C., et al., Cellulose, hemicelluloses, lignin and ash content of some organic materials and their suitability for use as paper pulp supplements. *Bioresource Technology*, 2007. 98(2): p. 296-301.

[86] Torrado, A.M., et al., Citric acid production from orange peel wastes by solid-state fermentation. *Brazilian Journal of Microbiology*, 2011. 42(394-409).

[87] Jekayinfa, S.O. and O.S. Omisakin, The Energy Potentials of Some Agricultural Wastes as Local Fuel Materials in Nigeria. *Agricultural Engineering International*, 2005. VII.

[88] Boumediene[1], M., et al., Characterization of two cellulosic waste materials (orange and almond peels) and their use for the removal of methylene blue from aqueous solutions. Maderas. *Ciencia y tecnología*, 2015. 17: p. 69-84.

[89] Unidad de Planeación Minero Energética, Resolución 355 de 2004, 2004.

[90] Aparcana, S.R. and A. Jansen, Estudios del valor fertilizante de los productos del proceso "Fermentación anaeróbica" para producción de biogás, German ProfEC.

[91] Deep Resource. Energy related conversion factors. 2012 [cited 2015 May 01]; Available from: https://deepresource.wordpress.com/2012/04/23/energy-related-conversion-factors/.

[92] Sustainable sanitation and water management. Direct use of biogas. [cited 2015 May 01]; Available from: http://www.sswm.info/content/direct-use-biogas.

[93] Pires, A.M.M., et al., Organic acids in the rhizosphere and phytoavailability of sewage sludge-borne trace elements. Pesquisa *Agropecuária B*rasileira, 2007. 42: p. 917-924.

[94] Clarke, W.P., et al., Digestion of waste bananas to generate energy in Australia. *Waste Management*, 2008. 28(3): p. 527-533.

[95] Kumar, P.K.R. and B.K. Lonsane, Gibberellic acid by solid state fermentation: Consistent and improved yields. *Biotechnology and Bioengineering,* 1987. 30(2): p. 267-271.

In: Citrus Fruits
Editor: Daphne Simmons

ISBN: 978-1-63484-078-1

Chapter 5

CITRUS RESIDUES AS SUPER-ADSORBENTS

Ioannis Anastopoulos[1] and George Z. Kyzas[2,3,*]

[1]Laboratory of Soils and Agricultural Chemistry, Department of Natural Resources and Agricultural Engineering, Agricultural University of Athens, Athens, Greece

[2]Department of Oenology and Beverage Technology, Technological Educational Institute of Kavala, Kavala, Greece

[3]Division of Chemical Technology, Department of Chemistry, Aristotle University of Thessaloniki, Thessaloniki, Greece

ABSTRACT

Water pollution is still a serious problem for the entire world. Adsorption technology is a promising process which based on fabrication of novel, cheap, non dangerous and highly sorptive materials for application in wastewater purification processes. *Citrus* species generally produced for the fresh consumption or the production of fruit juice but also have lot of application in medicine, food processing and agriculture sectors. This review collects information from published works about the alternative use of *Citrus* residues as efficient and promising adsorbents in clean water technology.

For this purpose, isotherm (Langmuir, Freundlich, etc.), kinetic (pseudo-first, -second order, etc.), thermodynamic (free energy Gibbs, enthalpy, entropy) and desorption-regeneration studies were discussed in detailed. Moreover, significant factors such as pH, agitation time, temperature, adsorbent dosage and initial dye concentration are also reported extensively.

Keywords: citrus peels, isotherms, kinetics, heavy metals, thermodynamics, modeling

* Corresponding author: George Z. Kyzas. University Campus (P.O. Box 116), Division of Chemical Technology, Department of Chemistry, Aristotle University of Thessaloniki, Thessaloniki 54124, Greece. E-mail: georgekyzas@gmail.com.

1. INTRODUCTION

Water pollution is still a momentous global issue that distresses the human population around the world. Every day, large amount of wastewater are generated by industrial, agricultural and domestic activities and are deposited to the land or water receptors. The latter is the common opinion causing the water pollution on a global scale. This is the reason why polluted wastewater must be depurated and returned to water receptors or to land.

There are many pollutants which contributing to environmental pollution such as dyes, radionuclides, phenolics, pesticides, etc. However, heavy metals are recognized as one of the most toxic groups which reach in food chain through the disposal of wastes to water receptors or land. Heavy metals are taxed in causing toxic effects, cancer and diseases because they cannot be degraded [1-3].

Various methods such as adsorption, coagulation, advanced oxidation, and membrane separation have been used for the removal of heavy metals from wastewaters. However, adsorption is one of the most effective processes of advanced wastewater treatment [4, 5]. Some of the widely used adsorbents materials especially for heavy metals removals from aqueous solutions are chitosans [6], nanoadsorbents [7], wastes from olive oil industry [8], algae [9], composts [10], zeolites [11], clay soils [12] and betonites and vermiculites [3]. In this study, the use of *Citrus* residues for heavy metal biosorption is discussed.

Citrus is a genus in the family Rutaceae and it is an important fruit crop with a total world production of 11.65 million metric tons. Colombia, China and Nigeria are on top (56.1% combined production of the world) among the citrus produces [13]. Citrus can be generally classified into the following categories: sweet oranges (most are *C. sinensis*), mandarins (*C. reticulata*), sour/bitter oranges (such as Seville, *C. aurantium*), lemons (*C. limon*), limes (*C. aurantifolia* and *latifolia*), grapefruit (*C. paradisi*) and pummelos (*C. grandis*), hybrids (e.g., tangelos, tangors, and limequats), and citrons (*C. medica*) [14]. World production of citrus by fruit type in 2010 was 56% oranges, 17% tangerines, clementines and mandarins, 11% lemons and limes, 6% grapefruit and pomelo and 10% other [15]. Generally, they are produced for the fresh consumption or the production of fruit juice. Moreover, the by-products generated following their industrial process are source of important bioactive compounds with potential for animal feed, manufactured foods and health care. They also consists of variety of phytonutrients such as alkaloids, flavonoids, tannins, phenols and saponins, which are closely related with health promotion and disease prevention [16]. Citrus species, also, pay attention for the production of essential oils which have plethora application in food, agriculture, pharmaceutical, cosmetology and sanitary areas [17].

Citrus biomass, has received a great deal of attention for decontamination of water by biosorption process. Their interest lies in the existence of functional groups such as sulfonic acids, carboxyl, that interact with the molecules of the contaminants found in the water resulting in their adsorption (physical or chemical) onto adsorbents surface. Another parameter which affects the adsorption is the surface area that taking place the adsorption process. High specific area increases the adsorption. A general sorbents production is as follows: citrus residues chopped, rinsed with distilled water, dried at constant temperature and then ground and sieve to a special particles size. Many scientists use different methods physical (heating/boiling) or chemical (alkali or acid solutions) to improve their adsorption capacity.

For first time, this review article summarizes and discusses *Citrus* residues (raw or treated) acting as biosorbents. Isotherm, kinetic, and equilibrium modeling were discussed in details. Moreover parameters which affect the biosorption process, such as the effect of solution pH, contact time, temperature and biosorbent's dose are also commented.

2. Modeling

In order to organize, establish and understand an adequate design model for the removal of pollutants from aqueous media, isotherm, kinetics and thermodynamic studies are essential basic prerequisites.

Adsorption isotherms models are extensively used to provide information about the amount of adsorbed ion by a certain biomass and about the interaction between the adsorbents and adsorbate [18, 19].

. In a recent review, Rangabhashiyam et al. [20] studied the use of two, three, four and five parameters isotherms models and concluded that in case of two parameters isotherm model, Langmuir and Freundlich models have the best fit. Langmuir isotherm assumes that all binding sites have equal affinity for the adsorbate, resulting to the formation of monolayer of adsorbed molecules [21] and Freundlich isotherm describes adsorption onto heterogeneous surfaces that provide adsorption sites of varying affinities [22]. The linear and non-lincar expressions of Langmuir and Freundlich isotherms are listed in Table 1.

Kinetics studies are useful tool to found the optimum condition for full-scale batch adsorption process [23]. Kinetic modeling reveal the mechanism of adsorption and potential rate-controlling steps such as mass transport or chemical reaction processes [23, 24]. The most common are the pseudo-first and the pseudo-second order kinetic equations. The linear and non linear forms of pseudo-first and pseudo-second-order kinetic models are presented in Table 1.

For better understanding the sorption mechanism, thermodynamic studies are basic requirements. They predict the spontaneity of adsorption and found the temperature range in which the adsorption was favorable or unfavourable [25]. The thermodynamic parameters for the adsorption process, Gibbs energy (ΔG^0), enthalpy of adsorption (ΔH^0) and entropy of adsorption (ΔS^0), are calculated by fitting data obtained by the adsorption experiments at different temperatures to the equations included in Table 1.

3. Citrus Residues

3.1. Citrus Sinensis (Orange)

Bhalerao et al. reports the removal of Cd^{2+} by using treated with 0.1 M HNO_3 orange rind [26]. Optimum adsorption condition was observed at pH 7 and 5 mg/ml biomass dose. Based on kinetic studies, the adsorption process was rapid and equilibration time was carried out after 90 min. FTIR analysis established that carboxyl, hydroxyl and carbonyl group was get involved in adsorption process. Mean free energy was equal to 1.118 kJ/mol, suggesting that the adsorption was a physical process.

Table 1. Isotherms, kinetics and thermodynamic equations

Expression	Equation form	Plot
Non linear Langmuir	$q_e = q_m \frac{b_L C_e}{1+b_L C_e}$	–
Linear Langmuir-1	$\frac{C_e}{q_e} = \frac{1}{q_m} C_e + \frac{1}{b_L q_m}$	$\frac{C_e}{q_e}$ vs C_e
Linear Langmuir-2	$\frac{1}{q_e} = \left(\frac{1}{b_L q_m}\right)\frac{1}{C_e} + \frac{1}{q_m}$	$\frac{1}{q_e}$ vs $\frac{1}{C_e}$
Linear Langmuir-3	$q_e = q_m - \left(\frac{1}{b_L}\right)\frac{q_e}{C_e}$	q_e vs $\frac{q_e}{C_e}$
Linear Langmuir-4	$\frac{q_e}{C_e} = b_L q_m - b_L q_e$	$\frac{q_e}{C_e}$ vs q_e
Non-linear Freundlich	$q_e = K_F C_e^{1/n}$	–
Linear Freundlich	$\ln q_e = \ln K_F + \frac{1}{n} \ln C_e$	$\ln q_e$ vs $\ln C_e$
Non-linear pseudo-first order kinetic	$q_t = q_e(1\text{-}\exp^{-k_1 t})$	–
Linear pseudo-first order kinetic	$\ln(q_e - q_t) = \ln q_e - k_1 t$	$\ln(q_e - q_t)$ vs t
Non-linear pseudo-second order kinetic	$q_t = \frac{k_2 q_e^2 t}{1+k_2 q_e t}$	–
Linear pseudo-second order kinetic	$\frac{t}{q_t} = \frac{1}{k_2 q_e^2} + \frac{1}{q_e} t$	t/q_t vs t
Gibbs	[a] $\Delta G^0 = -RT \ln b_L$ or [b] $\Delta G^0 = -RT \ln\left(\frac{q_e}{C_e}\right)$ [c] $\Delta G^0 = -RT \ln K$	–

Expression	Equation form	Plot
	[d] $\Delta G^0 = -RT \ln K_o$	
Van't Hoff	$\ln(b_L) = -\frac{\Delta H^0}{RT} + \frac{\Delta S^0}{R}$ or $\ln\left(\frac{q_e}{C_e}\right) = -\frac{\Delta H^0}{RT} + \frac{\Delta S^0}{R}$ $\ln(K) = -\frac{\Delta H^0}{RT} + \frac{\Delta S^0}{R}$ $\ln(K_0) = -\frac{\Delta H^0}{RT} + \frac{\Delta S^0}{R}$	$\ln(b_L)$ vs $\frac{1}{T}$ or $\ln\left(\frac{q_e}{C_e}\right)$ vs $\frac{1}{T}$ or $\ln(K)$ vs $\frac{1}{T}$ $\ln(K_0)$ vs $\frac{1}{T}$
Clausius Clapeyron	$\Delta H^0 = \frac{RT_1T_2}{T_2 - T_1}\left(\frac{\ln C_{e1}}{\ln C_{e2}}\right)$	–

C_e (mg/L) and q_e (mg/g) are the equilibrium liquid phase concentrations and amount of solute adsorbed at equilibrium, respectively; a: b_L in (L/mole) from Langmuir model; b: q_e/C_e is the ratio where q_e is adsorbed dye at concentration (mg/L) and C_e is the residual dye concentration in the solution (mg/L); c: $K = q_m X b_L$ calculated from Langmuir constants (q_m) is the maximum adsorption capacity in mg/g and b_L units in L/mg; d: K_0 can be evaluated by plotting ln (q_e/C_e) versus q_e by extrapolating to $q_e = 0$.

Table 2. List of models for adsorption isotherms, kinetics and thermodynamic for adsorption of metals on citrus peels

Adsorbent	Metal	Initial exp. conditions	Isotherm models	Kinetic models	q_m (mg/g)	Thermodynamics	Ref.
Citrus aurantifolia residues treated with $FeCl_3$	As^{5+}	t = 300-2280 min	-	ps1	0.475[a]	-	[28]
Grapefruit peel	As^{5+}	pH = 1-9; t = 0-1440 min T = 20-45°C	L, F	ps2, R, WM	37.77	-	[29]
Orange peel treated with 0.1 M HNO_3	Cd^{2+}	C_0 = 5-300 mg/L; pH = 2-10 t = 5-180 min	L	ps2	83.33	-ΔG, +ΔS, +ΔH	[26]
EDTA modified *Citrus sinensis* mesocarp	Cd^{2+}	C_0 = 10-50 mg/L; pH = 1-8 t = 20-120 min; T = 20-100°C	L	ps2	25.77	+ΔG, -ΔS, +ΔH	[30]

Table 2. (Continued)

Adsorbent	Metal	Initial exp. conditions	Isotherm models	Kinetic models	q_m (mg/g)	Thermodynamics	Ref.
Pomelo peels	Cd^{2+}	C_0 = 25-100 mg/L; pH = 1-6 t = 0-240 min; T = 25°C	L	ps2	21.83	-	[31]
Pretreated pomelo peels (isopropyl alcohol, sodium hydroxide)	Cd^{2+}	C_0 = 25-200 mg/L; pH = 2-6; t = 0-240 min; T = 25°C	L	-	26.88	-	[32]
Pretreated pomelo peels (isopropyl alcohol, sodium hydroxide, citric acid)	Cd^{2+}	C_0 = 25-200 mg/L; pH = 2-6; t = 0-240 min; T = 25°C	L	ps2	27.10	-	[32]
Orange peels	Ce^{3+}	C_0 = 50-350 mg/L; pH = 3-9; t = 120-420 min; dose = 0.05-0.35 g/L	L	–	71.40	-	[27]
Natural and chemically modified orange peels	Cu^{2+}	C_0 = 5-100 mg/L; pH = 2-10 t = 0-120 min	F	Ps2	(1.25-1.73)[a]	-	[33]
Orange peels	Cu^{2+}	C_0 = 1-100 mg/L; pH = 2-6.5 t = 30-90 min	L, F	ps2	14.75	-	[34]
Pomelo peels	Cu^{2+}	C_0 = 25-125 mg/L; pH = 3-6 t = 0-180 min; T = 25-45°C	L	ps2	19.7	-ΔG, +ΔS, -ΔH	[35]
Depectinated pomelo peels	Cu^{2+}	C_0 = 25-125 mg/L; pH = 3-6 t = 0-180 min; T = 25-45°C	L	ps2	21.1	-ΔG, +ΔS, -ΔH	[35]
Sour orange peels (*Citrus aurantium*)	Co^{2+}	C_0 = 5-50 mg/L; pH = 1-6 t = 30-150 min; T = 20-60°C dose = 0.5-3.0 g	F	-	-	-	[36]
Citrus reticulata waste biomass	Co^{2+}	C_0 = 25-800 mg/L; pH = 1-7 t = 15-1440 min; dose = 0.05-0.4 g T = 30-70°C	F	ps2	181.18	-	[37]
Orange peel treated with formaldehyde	Cr^{3+}	C_0 = 1-100 mg/L; pH = 2-3.5 t = 1-360 min	MCL	ps2	9.43	-	[38]
Citrus sinensis peels	Cr^{3+}	C_0 = 0.1- 0.3 mol/L; pH = 2-12 dose = 1-2 g; T = 30-60°C	F	-	-	±ΔG, -ΔS, -ΔH	[39]
Skin of orange peels	Cr^{6+}	C_0 = 40-100 mg/L; pH = 1-7 t = 20-90 min; T = 10-70°C	L	-	8.07	-	[40]
Orange peel treated with	Cr^{6+}	C_0 = 5-250 mg/L; pH = 1-8	L	ps2	10.74	-ΔG, +ΔS, +ΔH	[41]

Adsorbent	Metal	Initial exp. conditions	Isotherm models	Kinetic models	q_m (mg/g)	Thermodynamics	Ref.
0.1 M H_2SO_4		t = 10-180 min; dose = 1-15 g/L					
Mandarin peels	Cr^{6+}	C_0 = 50-500 mg/L; pH = 1-4 t = 0-480 min; T = 28-60°C dose = 1-5 g	-	-		-	[42]
Lemon skin	Cr^{6+}	C_0 = 10-30 mg/L; pH = 3-10 t = 10-80 min; T = 10-50°C dose = 0.025-0.2 g	L, F	ps1	13.69	-	[43]
Sweet lime skin	Cr^{6+}	C_0 = 10-30 mg/L; pH = 3-10 t = 10-80 min; T = 10-50°C dose = 0.025-0.2 g	L, F	ps1	6.53	-	[43]
Raw sweet lime	Ga^{3+}	C_0 = 40-200 mg/L; pH = 1-3 t = 0-300 min; T = 30°C	L	ps2	46.54	-	[44]
Alkali-treated sweet lime	Ga^{3+}	C_0 = 40-200 mg/L; pH = 1-3 t = 0-300 min; T = 30°C	L	ps2	76.26	-	[44]
Orange peel treated with formaldehyde	Fe^{3+}	C_0 = 3-300 mg/L; pH = 2-3.5 t = 1-360 min	MCL	ps2, Elv	18.20	-	[38]
Citrus reticulata waste biomass	Pb^{2+}	C_0 = 25-800 mg/L; pH = 1-5 t = 15-1440 min; dose = 0.05-0.4 T = 30-70°C	F	ps2	238.09	-	[37]
Pomelo peels	Pb^{2+}	C_0 = 25-100 mg/L; pH = 1-6 t = 0-180 min; T = 28°C	F	ps2	24.52[a]	-	[45]
EDTA modified *Citrus sinensis* mesocarp	Pb^{2+}	C_0 = 10-50 mg/L; pH = 1-8 t = 20-120 min; T = 20-100°C	L	ps2	13.94	+ΔG, -ΔS, +ΔH	[30]
EDTA modified *Citrus sinensis* mesocarp	Ni^{2+}	C_0 = 10-50 mg/L; pH = 1-8 t = 20-120 min; T = 20-100°C	L	ps2	33.44	+ΔG, -ΔS, +ΔH	[30]
Citrus Limettioides peel carbon	Ni^{2+}	pH = 2-12; t = 30-420 min dose = 0.05-3.5 g; T = 27-47°C	L	ps2	38.46	-ΔG, -ΔS, -ΔH	[46]
Citrus Limettioides seed carbon	Ni^{2+}	pH = 2-12; t = 30-420 min dose = 0.05-3.5 g; T = 27-47°C	L	ps2	35.54	-ΔG, -ΔS, -ΔH	[46]
Skin of orange peels	Zn^{2+}	C_0 = 40-100 mg/L; pH = 1-7 t = 20-90 min; T = 10-70°C	L	-	1.08	-	[40]

L: Langmuir; F: Freundlich; MCL: Multi component Langmuir; ps1: pseudo-first order; ps2: pseudo-second order; Elv: Elovich; R: Richenberg; WM: Weber-Morris; [a] Q_m values obtained from experimental results.

Orange peel was used for the adsorption of Ce^{3+} from aquatic system [27]. Maximum adsorption was noticed at pH 6 and 0.20 g/L of adsorbent dose. The kinetic study showed that an increase of the adsorbed amount was noticed till 4 h. A synopsis of best fitted isotherm and kinetic models, maximum adsorption capacities (obtained from Langmuir isotherm) and thermodynamic results are tabulated in Table 2.

Citrus sinensis peels were used for sequestration of Cr^{3+} by Ugbe et al. [39]. Increase of pH and temperature negatively affected the removal uptake and maximum adsorption was at pH 2 and 30°C. At low pH values, Cr^{3+} oxidised to form CrO_4^{2-} which is adsorbed by positively protonated adsorbent surface. It was observed that the removal of Cr^{3+} increase with increasing adsorbent dose. Gibbs free energy values were between -20 and 0 kJ/mol suggesting physisorption process.

The ability of orange rind treated with 0.1 M H_2SO_4 to adsorb Cr^{+6} from aqueous media was investigated by Poojari et al. [41]. Optimum adsorption conditions was observed at pH 2, 5 mg/ml adsorbent dose. Higher adsorbent dose leads to the reduction of metal uptake due to partial aggregation of adsorbent contributing to the decrease in effective surface area. Equilibrium was reached in 150 min. Based on estimated mean free energy the adsorption was found to be physical in nature.

Natural (OP-Natural) and chemically modified orange peels (OP-NaOH, OP-Methanol, OP-H_2SO_4, OP-Formaldehyde, OP-Acetic anhydride) were tested for their ability to adsorb Cu(II) ions from water media by Khalfaoui and Meniai [33]. pH was found to affect significant the adsorption process and optimum adsorption was observed at pH 5-6. Chemical treatments via NaOH, H_2SO_4 and methanol (esterification), were found to increase the adsorption percentage from 75% (natural OP) to 91-99%, suggested the success of modification process to improve the adsorption capacity. Adsorption was rapid and equilibrium was achieved in 5 min for OP-NaOH, OP-Methanol, OP-H_2SO_4 and 10 min for OP-Natural.

Orange peel was also examined for the removal of Cu^{2+} from aqueous solution by khan et al. [34]. Adsorption increased with the increase of pH and 6.5 pH value was chosen as optimum pH value for adsorption experiments. At pH > 6.5, Cu precipitates as $Cu(OH)_2$, hindering the biosorption process. Kinetic studies were conducted and showed that the adsorption was fast and equilibrium was attained at 60 min.

EDTA modified *Citrus sinensis* mesocarp was used as adsorbent for its ability to sequestrate Pb^{2+}, Cd^{2+} and Ni^{2+} ions from aqueous solution [30]. The results showed that at low pH values, the hydrogen ions act competed thus maximum adsorption was observed at pH values 5, 6 and 7 for Ni^{2+}, Cd^{2+} and Pb^{2+}, respectively.

Adsorption was found to increase as contact time increased and equilibrium was attained in 100 min. The increase of temperature was positively affected the adsorption capacity due to the fact that at high temperatures the thickness of the outer surface reduces and the number of pores increases resulting in more efficiently attachment of the metal ion onto the surface of adsorbent.

The utilization of pretreated via formaldehyde orange peel as adsorbent for the uptake of Fe(III) and Cr(III) in single and binary systems was examined by Lugo-Lugo et al. [38]. The increase of pH lead to the increase of uptake amount and maximum adsorption was achieved at pH 4 due to the decrease of H^+ ions and the formation of metal hydroxide complexes. To avoid iron precipitation, pH 3 was selected for the rest of the experiments. In single metal system, for Cr(III) and Fe(III) equilibrium was found at 260 and 120 min, respectively. On the contrary, in binary system, 120 min was observed to be sufficient for both metals to equilibrate.

In binary systems removal uptake was found to decrease from 7.6 to 4.79 mg/g and 17.4 to 8.96 mg/g for Cr(III) and Fe(III), respectively. This is the reason why both metals have antagonism character, for the same binding sites. Hydroxyls groups on the carboxylic acids, based on FTIR spectra, recognized as the active binding sites.

Another research team was examined the removal of Cr^{6+} and Zn^{2+} by skin of orange peel [40]. Optimum removal of Cr and Zn ions was observed at pH 3. The increase of adsorption capacity was achieved by raising the temperature up to 30°C while the opposite results were observed with further increase. It is explained that at high temperatures the entropy increases as a result the reduction of the stability of metals ions; finally lower amount of heavy metals is available to adsorb. Adsorbent dose of 1 g leads to highest uptake percentage due to greater availability of exchange binding sites or larger surface area. Kinetic studies showed that 30 min was sufficient for equilibration.

3.2. Citrus Paradisi (Grapefruit), Citrus Grandis (Pomelo), Citrus Reticulata (Mandarin), Citrus Aurantium (Sour Organge)

Batch experiments were applied to examine the uptake of As^{5+} from wastewater [29]. Maximum adsorption was found at pH 4 and 45°C. Kinetic studies revealed that equilibrium was attained in 120 min.

The mechanism of the adsorption indicated that both surface adsorption and intra-particle diffusion get involved in adsorption process. FTIR characterization affirms the involvement of carbonyl and hydroxyl in adsorption process. Application of grapefruit peel in real wastewater proved to be good adsorbent for the removal of As^{5+} lead to water up to safe limits for As contamination, as proposed from WHO.

Tasaso [35] utilized pomelo and depectinated pomelo peels as adsorbents for the removal of Cu^{2+} from aqueous solutions. Maximum adsorption was at pH 4 and equilibrium was achieved at 60 min. Increase of temperature from 25 to 45°C, negatively affected the adsorption of Cu^{2+}. FTIR spectra (before and after adsorption) indicated significant shift of COO^- groups, indicating the involvement in adsorption process.

Pomelo peels were also used for adsorption of Pb^{2+} [45]. Biosorption was found to be rapid and equilibrium was reached in 90 min. Optimum pH value was 4 with maximum adsorption percentage 96.79%.

Raw pomelo peels (RPP) and its modified forms (PPI: washed with isopropyl alcohol, PPIS: washed with isopropyl alcohol and sodium hydroxide, PPIC: washed with isopropyl alcohol and citric acid, PPISC: washed with isopropyl alcohol, sodium hydroxide and citric acid) were used as efficient adsorbents to remove Cd^{2+} [31, 32].

Maximum adsorption was achieved at pH 5. For all adsorbents, kinetic studies revealed a rapid biosorption process in the first 20-30 min and after 60 min a plateau was noticed. Maximum adsorption capacity was observed by PPISC adsorbent at pH 5.

The use of mandarin peel was examined for Cr^{6+} removal from aqueous media using batch trial approached [42]. At 28°C, optimum pH and time conditions was 1 and 6.5 h, respectively. Increase of temperature up to 60°C positively affected the adsorption uptake capacity. At 5 g of adsorbent and 35 min, 100% removal of Cr^{6+} was observed. The authors explained that higher amount of adsorbent dose contributing to more active binding sites thus facilitating the adsorption process. Desorption studies were conducted by using 0.1 and 0.5 N NaOH and the results showed that desorption percentage was 83 and 62.86% with 0.1 N and 0.5 N NaOH, respectively.

Citrus reticulata was also used for the adsorption of Co^{2+} and Pb^{2+} from water media [37]. Experiments were carried out as function of pH, adsorbent dose, adsorbent size, contact time and temperature. Pb^{2+} and Co^{2+} maximum adsorption was found at pH 5, 0.1 g dose, 0.31 mm size, and pH 7, 0.05 g dose, 0.5 mm size, respectively. Pb^{2+} removal was temperature independent while Co^{2+} uptake positively affected by increasing of temperature, giving maximum adsorption at 60°C. Among tested modifications of raw biomass, sodium hydroxide and simply heated pre-treatments gave maximum capacity for Pb^{2+} (83.77 mg/g) and Co^{2+} (95.55 mg/g), respectively.

Sour orange was investigated as potential adsorbent for the removal of Co from aqueous solution [36]. Optimum adsorption condition were found at pH 2, 30°C, adsorbent dose 2 g and contact time 90 min. Desorption studies were carried out using sodium chloride, EDTA, hydrochloric acid, ammonium chloride and calcium chloride. NH_4Cl gave the highest desorption efficiency 95.12%. FTIR analysis before and after adsorption demonstrated significant changes to hydroxyl and carboxyl groups suggested the potential participation in adsorption process.

3.3. Citrus Aurantifolia (Key Lime), Citrus Limettioides (Sweet Lime), Citrus Lemon (Limon)

Marín-Rangel et al. [28] utilized *Citrus aurantifolia* residues treated with $FeCl_3$ for the adsorption of As^{5+}. Chemical composition analysis showed that the adsorbent consisted of lignin, hemicellulose, pectin, cellulose, carbohydrates and proteins. SEM studies indicated the heterogeneous surface of adsorbent with smooth and rough areas. Kinetic studied concluded that equilibrium was reached at 24 h, giving uptake rate of 58.28%.

Activated carbon from *Citrus Limettioides* peel (CLPC) and seed (CLSC), respectively, were fabricated and tested for its ability to adsorb Ni^{2+} [46]. Optimum pH value and equilibrium time was determined at pH 4-7 and 4 h, respectively. Adsorbent dose of 1.5 g/L gave the maximum adsorption capacity (99%) for both activated carbons. Regeneration studies were achieved by using 0.7 N HCl and after 5 adsorption-desorption cycles the uptake removal was decreased from 96.20 to 80.70% and from 94.50 to 78.90%, for CLPC and CLSC, respectively. CLPC and CLSC were also applied in real wastewaters containing Ni^{2+} ions, giving satisfactorily adsorptive results (maximum removal of 98% (± 0.5) at adsorbent dose of 3 g/L).

Sweet lime and lemon skin were used for the adsorption of Cr^{6+} from water media by Phadtare and Patil [43]. pH was found to control strongly the adsorption process and maximum adsorption was obtained at pH values 4-6. Increase of temperature negatively affected the adsorption removal demonstrated the exothermic nature of adsorption process. A decrease of adsorbent size and increase of adsorbent dose was found to enhance the adsorption procedure. Initial concentration had not have significant impact on the removal of Cr^{6+} by both adsorbents.

Gondhalekar and Shukla [44] utilized raw (RCP) and alkali treated (ACP) *Citrus limetta* peels for their adsorption capacity for Ga^{3+}. FTIR spectra showed that the alkali pretreatment converted ester functional groups to carboxylic acid groups. Batch studies were carried out and the results showed that optimum adsorption was obtained at pH 3 and equilibrium was attained in 180 min. Maximum adsorption capacity was found to be enhanced by alkali pretreatment (RCP: 46.54 mg/g, ACP: 76.26 mg/g), indicated the success of pretreatment. After 3 adsorption-desorption cycles the adsorption capacity decreased from 58.98 to 37.08% and 68.51 to 55.46%, for RCP and ACP, respectively.

Table 3. Thermodynamic parameters for studied adsorbents

Adsorbent	Metal	T (K)	ΔG^0 (kJ/mole)	ΔH^0 (kJ/mole)	ΔS^0 (kJ/mole K)	Ref.
Orange peel treated with 0.1 M HNO_3	Cd^{2+}	303	-0.77	13.32	0.046	[26]
		313	-0.96			
		323	-1.68			
Orange peel treated with 0.1 M H_2SO_4	Cr^{6+}	293	-0.21	8.84	0.030	[41]
		303	-0.81			
		313	-0.84			
		323	-1.19			
Pomelo peels	Cu^{2+}	298	-5.38	-32.18	0.09	[35]
		308	-4.19			
		318	-3.49			
Depectinated pomelo peels	Cu^{2+}	298	-5.40	-24.89	0.08	[35]
		308	-4.72			
		318	-4.02			
Orange peel treated with 0.1 M H_2SO_4	Cr^{+6}	293	-0.21	8.84	0.030	[41]
		303	-0.81			
		313	-0.84			
		323	-1.19			
Citrus Limettioides peel carbon	Ni^{2+}	300	-6.09	-21.88	-0.053	[46]
		310	-5.67			
		320	-5.06			
Citrus Limettioides seed carbon	Ni^{2+}	300	-5.49	-19.09	-0.045	[46]
		310	-4.90			
		320	-4.60			

4. Isotherm, Kinetic and Thermodynamic Analysis

In most cases, experimental data well described from Langmuir model, pseudo-second-order kinetic and the sorption process is spontaneous (Table 2). Thermodynamic studies were calculated by many ways, as described in Table 1. In the majority of adsorption studies, authors, without explanation use one of these equation and plots and the final decision of the most suitable plot was proved by R^2 values.

Thermodynamic parameters by using citrus residues for metals adsorption are tabulated in Table 3. As we can see, sorption data from three to four temperatures were used in a range 293-323. Thermodynamic studies showed that the adsorption process was spontaneous ($\Delta G^0 < 0$) and endothermic ($\Delta H^0 > 0$) or exothermic ($\Delta H^0 < 0$).

Conclusion

Adsorption process is still a promising technology in clean water science. Among tested adsorbents, citrus based biomass is noticeable attractive adsorbents due to its higher adsorption capacity for target heavy metal. Adsorption was found to be controlled by initial concentration, adsorbent dose, contact time, solution pH and temperature. Among tested isotherm and kinetic models, Langmuir and pseudo-second order were found to express well the adsorption process.

Despite the number of published adsorption studies, there is a lack of information regarding the behavior of citrus adsorbents in multi-metal systems and the application in real wastewaters. Mechanism of adsorption is a very complicate issue and must be accompanied by multi-faceted studies and not only numerically by modeling equations.

Based on the fact that most of studies focus only on laboratory experiments, future work must give emphasis to estimate the cost of fabrication and the application of citrus based adsorbents in industry effluents.

References

[1] Dousis, P., Anastopoulos, I., Gasparatos, D., Ehaliotis, C., Massas, I. (2012). Effects of time and glucose-c on the fractionation of Zn and Cu in a slightly acidic soil. *Commun. Soil Sci. Plant Anal.* 44, 722-732.

[2] Farooq, U., Kozinski, J. A., Khan, M. A., Athar, M. (2010). Biosorption of heavy metal ions using wheat based biosorbents - A review of the recent literature. *Bioresour. Technol.* 101, 5043-5053.

[3] Malamis, S., Katsou, E. (2013). A review on zinc and nickel adsorption on natural and modified zeolite, bentonite and vermiculite: Examination of process parameters, kinetics and isotherms. *J. Hazard. Mater.* 252-253, 428-461.

[4] Kyzas, G. Z., Fu, J., Lazaridis, N. K., Bikiaris, D. N., Matis, K. A. (2015). New approaches on the removal of pharmaceuticals from wastewaters with adsorbent materials. *J. Mol. Liq.* 209, 87-93.

[5] Vijayaraghavan, K., Balasubramanian, R. (2015). Is biosorption suitable for decontamination of metal-bearing wastewaters? A critical review on the state-of-the-art of biosorption processes and future directions. *J. Environ. Manage.* 160, 283-296.

[6] Kyzas, G. Z., Bikiaris, D. N. (2015). Recent modifications of chitosan for adsorption applications: A critical and systematic review. *Mar. Drugs* 13, 312-337.

[7] Kyzas, G. Z., Matis, K. A. (2015). Nanoadsorbents for pollutants removal: A review. *J. Mol. Liq.* 203, 159-168.

[8] Anastopoulos, I., Massas, I., Ehaliotis, C. (2015). Use of residues and by-products of the olive-oil production chain for the removal of pollutants from environmental media. A review of batch biosorption approaches. *J. Environ. Sci. Health, Part A* 50, 677-718.

[9] Anastopoulos, I., Kyzas, G. Z. (2015). Progress in batch biosorption of heavy metals onto algae. *J. Mol. Liq.* 209, 77-86.

[10] Anastopoulos, I., Kyzas, G. Z. (2015). Composts as biosorbents for decontamination of various pollutants: A review. *Water, Air, and Soil Pollution* 226, Article ID 61.

[11] Anastopoulos, I., Ioannou, D., Kallianou, C. (2012). Removal of heavy metals from aqueous solutions through natural Greek clay. Selectivity order and isotherms studies. *Agrochimica* 56, 98-111.

[12] Olaofe, O., Olagboye, S. A., Akanji, P. S., Adamolugbe, E. Y., Fowowe, O. T., Olaniyi, A. A. (2014). Kinetic studies of adsorption of heavy metals on clays. *Int. J. Chem.* 7, 48-54.

[13] Khan, S. A., Ahmad, R., Asad, S. A., Shahzad, M. (2014). Citrus flavonoids: Their biosynthesis, functions and genetic improvement. In: *Citrus molecular phylogeny, antioxidant properties and medicinal uses*, New York, US, Nova Science Publishers, pp. 31-51.

[14] Turner, T., Burri, B. J. (2013). Potential nutritional benefits of current citrus consumption. *Agriculture* 3, 170-187.

[15] FAOSTAT (2012). Production crops. Food and agriculture organization of the United Nations. http://faostat3.fao.org/home/index.html# VISUALIZE (accessed on 5 October 2012).

[16] Okwu, D. E. (2008). Citrus fruits: a rich source of phytochemicals and their roles in human health. *International Journal of Chemical Sciences* 6, 451-471.

[17] Palazzolo, E., Laudicina, V. A., Germanà, M. A. (2013). Current and potential use of citrus essential oils. *Curr. Org. Chem.* 17, 3042-3049.

[18] Bharathi, K. S., Ramesh, S. T. (2013). Removal of dyes using agricultural waste as low-cost adsorbents: a review. *Appl. Water Sci.* 3, 773-790.

[19] Ramesh, A., Lee, D. J., Wong, J. W. C. (2005). Thermodynamic parameters for adsorption equilibrium of heavy metals and dyes from wastewater with low-cost adsorbents. *J. Colloid Interface Sci.* 291, 588-592.

[20] Rangabhashiyam, S., Anu, N., Giri Nandagopal, M. S., Selvaraju, N. (2014). Relevance of isotherm models in biosorption of pollutants by agricultural byproducts. *J. Environ. Chem. Eng.* 2, 398-414.

[21] Langmuir, I. (1918). The adsorption of gases on plane surfaces of glass, mica and platinum. *J. Am. Chem. Soc.* 40, 1361-1403.

[22] Freundlich, H. (1906). Over the adsorption in solution. *Z. Phys. Chem.* 57, 385-470.

[23] Febrianto, J., Kosasih, A. N., Sunarso, J., Ju, Y.-H., Indraswati, N., Ismadji, S. (2009). Equilibrium and kinetic studies in adsorption of heavy metals using biosorbent: A summary of recent studies. *J. Hazard. Mater.* 162, 616-645.

[24] Park, D., Yun, Y.-S., Park, J.-M. (2010). The past, present, and future trends of biosorption. *Biotechnology and Bioprocess Engineering* 15, 86-102.

[25] Doke, K. M., Khan, E. M. (2013). Adsorption thermodynamics to clean up wastewater; critical review. *Reviews in Environmental Science and Biotechnology* 12, 25-44.

[26] Bhalerao, S. A., Poojari, A. C., Maind, S. D. (2015). Biosorption studies of cadmium(II) ions from aqueous solutions onto orange rind (*Citrus sinensis* L. Osbeck). *Octa Journal of Environmental Research* 3, 28-40.

[27] Varshini, C. J. S., Das, N. (2014). Screening of biowaste materials for the sorption of cerium(III) from aqueous environment. *Research Journal of Pharmaceutical, Biological and Chemical Sciences* 5, 402-408.

[28] Marín-Rangel, V. M., Cortés-Martínez, R., Cuevas Villanueva, R. A., Garnica-Romo, M. G., Martínez-Flores, H. E. (2012). As(V) biosorption in an aqueous solution using chemically treated lemon (*Citrus aurantifolia swingle*) residues. *J. Food Sci.* 77, T10-T14.

[29] Khaskheli, M. I., Memon, S. Q., Parveen, S., Khuhawar, M. Y. (2014). Citrus paradisi: An effective bio-adsorbent for arsenic(V) remediation. *Pakistan Journal of Analytical and Environmental Chemistry* 15, 35.

[30] Jimoh, T. O., Bankole, M. T., Muriana, M., Abdullahi, F. O. M.-E. J. o. S. R., 13(5), 585-593. (2013). Sequestration of Pb(II), Cd(II) and Ni(II) ions from aqueous solution using EDTA modified *Citrus sinensis* mesocarp. *Middle-East Journal of Scientific Research* 13, 585-593.

[31] Saikaew, W., Kaewsarn, P., Saikaew, W. (2009). Pomelo peel: agricultural waste for biosorption of cadmium ions from aqueous solutions. *World Acad. Sci. Eng. Technol.* 56, 287-291.

[32] Saikew, W., Kaewsarn, P., 2010, 'Pretreated pomelo peel as biosorbent of cadmium ion from aqueous solution,' *The 8th Asian-Pacific Regional Conference on Practical Environmental Technologies (APRC2010)* Ubon Ratchathani University, Thailand.

[33] Khalfaoui, A., Meniai, A. H. (2012). Application of chemically modified orange peels for removal of copper(II) from aqueous solutions. *Theor. Found. Chem. Eng.* 46, 732-739.

[34] Khan, S., Farooqi, A., Danish, M. I., Zeb, A. (2013). Biosorption of copper(II) from aqueous solution using citrus sinensis peel and wood sawdust: Utilization in purification of drinking and waste water. *Int. J. Res. Rev. Appl. Sci.* 16, 297-306.

[35] Tasaso, P. (2014). Adsorption of copper using pomelo peel and depectinated pomelo peel. *Journal of Clean Energy Technologies* 2, 154-157.

[36] Sattar, J. A. A. (2013). Toxic metal pollution abatement using sour orange biomass. *Journal of Al-Nahrain University* 16, 56-64.

[37] Bhatti, H., Bajwa, I., Hanif, M., Bukhari, I. (2010). Removal of lead and cobalt using lignocellulosic fiber derived from *Citrus reticulata* waste biomass. *Korean J. Chem. Eng.* 27, 218-227.

[38] Lugo-Lugo, V., Barrera-Díaz, C., Ureña-Núñez, F., Bilyeu, B., Linares-Hernández, I. (2012). Biosorption of Cr(III) and Fe(III) in single and binary systems onto pretreated orange peel. *J. Environ. Manage.* 112, 120-127.

[39] Ugbe, F. A., Pam, A. A., Ikudayisi, A. V. (2014). Thermodynamic properties of chromium(III) ion adsorption by sweet orange (*Citrus sinensis*) peels. *American Journal of Analytical Chemistry* 5, 666-673.

[40] Ekpete, O. A., Kpee, F., Amadi, J. C., Rotimi, R. B. (2010). Adsorption of chromium(VI) and zinc(II) Ions on the skin of orange peels (*Citrus sinensis*). *Journal of Nepal Chemical Society* 26, 31-39.

[41] Poojari, A. C., Maind, S. D., Bhalerao, S. A. (2015). Effective removal of Cr(VI) from aqueous solutions using rind of orange (*Citrus sinensis*), (L.) Osbeck. *Int. J. Curr. Microbiol. App. Sci.* 4, 653-671.

[42] Acosta-Rodríguez, I., Coronado-Quintero, E., Cárdenas-González, J. F., Tovar-Oviedo, J., Martínez-Juárez, V.-M. (2013). Hexavalent chromium removal *Citrus reticulata shell*. *Journal of Natural Sciences* 1, 29-39.

[43] Phadtare, M. J., Pati, S. T. (2015). Removal of heavy metal from industrial wastewater. *International Journal of Advanced Engineering Research and Studies,* April-June, 4-8.

[44] Gondhalekar, S. C., Shukla, S. R. (2014). Recovery of Ga(III) by raw and alkali treated citrus limetta peels. *International Scholarly Research Notices* 2014, Article ID 968402.

[45] Andres, K. A. M., Bawalan, J. B., Galang, N. K. P. (2012). Use of *Citrus grandis (L.) Osbeck (Fam. Rutaceae)* peels waste material as a biosorbent for lead contaminated water. *International Journal of Chemical and Environmental Engineering* 3, 80.

[46] Sudha, R., Srinivasan, K., Premkumar, P. (2015). Removal of nickel(II) from aqueous solution using Citrus Limettioides peel and seed carbon. *Ecotoxicol. Environ. Saf.* 117, 115-123.

In: Citrus Fruits
Editor: Daphne Simmons

ISBN: 978-1-63484-078-1

Chapter 6

CITRUS GENETIC IMPROVEMENT: NEW CITRUS HYBRIDS FROM BREEDING PROCEDURES AND EVALUATION OF THEIR GENETIC AND PHYTOCHEMICAL ASPECTS

Edoardo Napoli*[1]*, Giuseppe Ruberto*[1,*]*,
***Loredana Abbate*[2]*, Francesco Mercati*[2]**
***and Sergio Fatta Del Bosco*[2]**
[1]CNR - Istituto di Chimica Biomolecolare, Catania, Italy
[2]CNR - Istituto di Bioscienze e BioRisorse, Palermo, Italy

ABSTRACT

The Citrus genetic improvement is obtained throughout the application of several breeding procedures of extant species. Main aims of such breeding approaches are to obtain seedless fruits with easily removable peel, optimal size, excellent and original organoleptic characters, and possibly fruits endowed with precocious or late ripening. Citrus fruits and some of their transformation products, such as juices, fall in the large category of the functional foods owing to their content of important secondary metabolites defined nutraceutical components, whose beneficial effects on the human health are continuously evidenced. In this context the aim of the breeding processes is to obtain new varieties with an increased amount of nutraceutical components. Besides these characters mainly associated to the new fruits, other important agronomic and economic aspects concern the production of plants with high productivity and improved resistance against biotic and abiotic stresses.

On these bases, our groups have focused the research activity in the genetic improvements of high quality cultivars and the production of new citrus fruits, namely hybrids. In particular, our interest, has been addressed to the study of the chemical composition (mainly polyphenols from juices and peel essential oils) of new Citrus hybrids, with the aim of an exhaustive phytochemical characterization and, possibly, the

* Corresponding author: Giuseppe Ruberto; CNR – Istituto di Chimica Biomolecolare; Via Paolo Gaifami, 18; 95126 Catania; Italy; Tel.: 0039 095 7338347; Fax: 0039 095 7338310; e-mail: giuseppe.ruberto@icb.cnr.it.

evaluation of these new fruits for their introduction into the fresh market and into the industrial chain of transformation.

The new hybrids have been obtained through somatic hybridization by protoplast fusion. This technique, enabling to combine fully or partially, nuclear and cytoplasmic genomes at the interspecific and intergeneric levels, allows to widen the gene pool and to increase the genetic diversity of a species, circumventing the naturally occurring sexual incompatibility barriers (nucellar polyembryony, long juvenility and pollen/ovule sterility). Following this approach, our breeding program has given rise to dozens of somatic hybrid and cybrids that are now being evaluated for their agronomic and productive characters.

A wide description of the different adopted breeding strategies and a summary of the phytochemical analyses of the new varieties obtained in these last years will be given.

INTRODUCTION

During the last century, particularly in the second half, the genetic progress has been extremely effective for many crops, for which improved cultivars significantly increased the yield potential [1-3]. In Citrus, the third most relevant commercial fruit crop in the world after apple and banana [4], huge efforts have been spent in order to improve the agronomic important traits, such as seedlessness, fruit yield, quality and characteristics, stress and disease tolerance, and several new citrus varieties have been entered into the international market in the last 20-30 years. Introduction of new traits has been based mainly on bud spot selection and sexual crossing within or between different genotypes and closely related species. However, introgression of desirable and agronomically interesting genes through conventional breeding approaches in Citrus is a very arduous task and, very often, the results are not abreast with the efforts. Indeed, various unavoidable biological limitations greatly hinder genetic manipulation and cultivar improvement through sexual hybridization: sexual (cross and self) incompatibility, nucellar embryogenesis and polyembriony, male and/or female sterility, different bloom phases, long juvenility and long generation time, large plant size [5]. All these factors, combined together with the lack of sufficient knowledge about the patterns of inheritance of horticultural traits, negatively affect breeding efforts in citrus cultivar development. The final bottleneck is the long time necessary to test the new genotypes: long (12-20 years) and multi-locus field trials are necessary to determine if yield, fruit quality, fruit holding capacity, disease tolerance, graft compatibility and survivability of a selected improved citrus cultivar are adequate and stable, before supplying it to the market.

BIOTECHNOLOGICAL APPROACHES TO CITRUS IMPROVEMENT

A range of biotechnological approaches available, such as somatic hybridization and molecular genetics, can help to circumvent some of the biological barriers associated with the reproductive biology of citrus.

A. Somatic Hybridization

The enormous advantage of tissue culture is the possibility the technique offers to plant breeders for the exploitation of genetic diversity within a short period, enabling the *in vitro* selection, the conservation and the exchange of germplasm. Following *in vitro* tissue culture systems, citrus plant regeneration has been reported by culturing different tissues, such as stem, epicotyl and root segments, nucellar embryos, ovules, anthers, pollen, stigma, style, juice vesicles, internodal stem segments [6-18]. However, the growth of plant cells and tissues *in vitro* and their regeneration into plants is an asexual process involving only mitotic division and the result is the regeneration of genetically uniform plants. The occurrence of variation during the regeneration process is generally null or very limited, therefore the tissue culture is not expected to provide a useful source of variation for plant breeding.

In order to enhance the variability of Citrus and to circumvent the limitations due to their complicated reproductive biology, the most effective tool is somatic hybridization by protoplast fusion [19-21]. Protoplasts are totipotent naked cells (the cell wall has been removed following enzymatic treatment) that have the capability to dedifferentiate, regenerate a new wall, re-enter the cell-cycle, go through mitotic divisions, regenerate various organs and, finally, whole plants. Somatic hybridization via protoplast fusion is an additive process capable to capture the genetic diversity of the gene pools by combining (fusing) the nuclear, chloroplast and mitochondrial genomes of desired parental protoplasts in a novel arrangement, therefore creating different and unobtainable (no meiotic segregation occurs during somatic hybridization process) homokaryon or heterokaryon types. The potential heterozygosity is extremely large depending on cumulative allelic differences between the contributing parents. Somatic hybrids can be classified into three types: symmetric somatic hybrids, asymmetric somatic hybrids and cybrids (cytoplasmic hybrids). Symmetric somatic hybridization is defined as the combination of nuclear and cytoplasmic genetic information from both parental species. Asymmetric somatic hybridization is incomplete, with the loss of some cytoplasmic or nuclear DNA, and this type of hybridization is used to introduce fragments of the nuclear genome from one donor species into the intact genome of another recipient species. Cybrids express only one parental nuclear genome and either the cytoplasmic genome of the other (non-nuclear) parent or that of a combination of both parental species.

Citrus and related genera are, compared to other genera, among the plants for which the protoplast-to-plant systems are more effective.

The initial purpose of somatic hybridization in citrus was to generate symmetric allotetraploid (amphidiploid) somatic hybrids containing the complete nuclear and cytoplasmic genomes of both parents, for use in cultivar as well as in rootstock improvement programs. Somatic hybrids may be directly used as tetraploid scions [19, 22, 23] or, if fertile, involved in interploid crosses to produce seedless triploid cultivars [24, 25]. On the other hand, somatic hybrids may be used in rootstock improvement in an effort to combine all genes related to biotic and abiotic tolerance of the two parents, for superior adaptation and productivity. In citrus, tetraploidy is known to affect phenotypic traits (cell size, stomata density, organellar composition, leaf thickness, root morphology, growth and fruit quality), physiological characters (reduction of growth rate and lowering rates of whole-plant transpiration) and agronomical features (higher tolerance to salt stress and better adaptation to water deficit) [26-28].

The first intergeneric allotetraploid somatic hybrid in citrus was produced thirty years ago between *Citrus sinensis* and *Poncirus trifoliata* [29]; since then, somatic hybridization has become an integral part of the citrus variety and rootstock improvement programs worldwide. By using the fusion model of "diploid embryogenic protoplasts + diploid mesophyll-derived protoplasts," to date, more than 250 different parental combination of intra and inter-specific and intergeneric somatic hybrid of citrus have been produced in efforts to improve scion agronomic performance, to enhance rootstock tolerance to diseases and to improve soil/environmental adaptability [24, 25]. Among the different objectives of somatic hybridization in citrus improvement, one of the major goals is the obtainment of plants bearing fruits of original and good sensory characteristics. The idea is that the large extent of genomic arrangement and recombination following ploidy manipulation during somatic hybridization process may have a deep impact on the chemical composition of somatic hybrid fruits, aiming to a presence of distinctive and original traits in their phytochemical characters.

The aromatic quality of Citrus allotetraploid somatic hybrids has been studied in the last 10-15 years [30-34]. It is generally difficult to establish an inheritance mechanism related to the biosynthetic accumulation of secondary metabolites and volatile components in Citrus somatic hybrids. Some studies suggest that complex forms of dominance originating from the genome of one of the two parents determine, to some extent, the biosynthesis pathways of some of the volatile compounds [32]. Other studies show an intermediate essential oil profile in the somatic hybrids with respect to those of the parents, suggesting that both parents may have contributed to the composition of volatile components, even though a predominant influence of one of the two parents may be detected [30]. The production of essential oil in the fruit peel of some interspecific somatic hybrids in which phytochemical composition exhibit original features with respect to those of both parents, prompted some authors to claim that Citrus somatic hybrids do not retain their parental traits [30, 33]. Investigations focused on the inheritance of characters involved in fruit quality in a Citrus interspecific allotetraploid somatic hybrid and its diploid donor parents using quantitative RT-PCR to analyze gene expression differences, have revealed that the genes controlling the biosynthetic pathways of the compounds studied (sugars, acids and carotenoid in the juice and aromatic compounds in the peel oil) are not inherited in an additive fashion in the allotetraploid hybrid and may be subject to dosage effects, likely over-dominance, co-dominance and other complex interactions in gene expression regulation [35, 36].

The presence in citrus somatic hybrids of distinctive phytochemical profiles is, however, a demonstration of the high potential of somatic hybridization in cultivar improvement. However, symmetric fusion, combining entire genomes of both parents, may contain many unwanted nuclear encoded agronomic traits derived from the parental species or from the wild species. Moreover, plants recovered from distant recombination are often morphologically abnormal, may show weakness or uncontrolled genomic instabilities, harbor fruits with thick and irregular rind like that of the typical tetraploid, thus hampering their direct use in crop genetic improvement [37, 38]. It has also been reported that many hybrids derived from symmetric fusions are sterile or characterized by low fertility, owing to disorders in the formation of functional gametes caused by imbalance of recombinant genome [39]. Therefore, no allotetraploid somatic hybrids have been extensively used in citrus commercial orchards, so far.

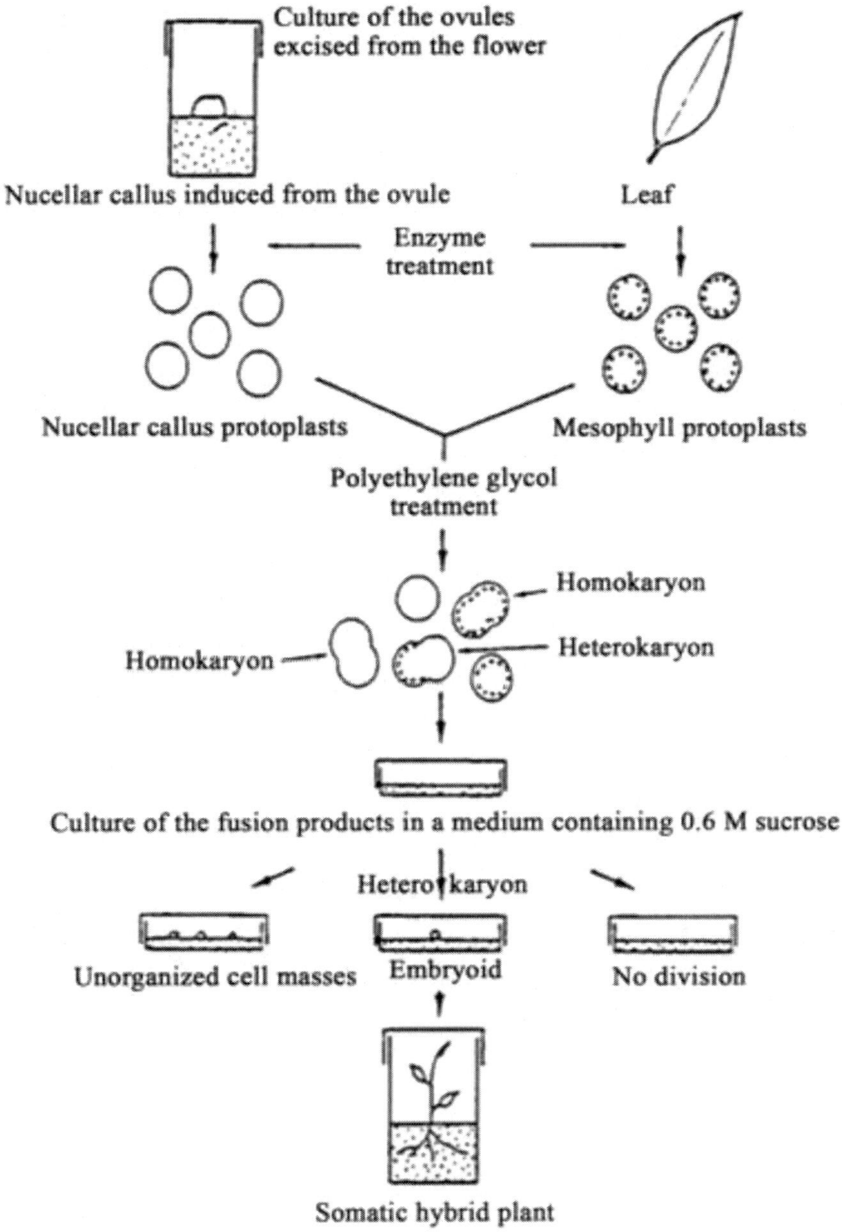

Figure 1. A schematic framework for the obtaining of *Citrus* somatic hybrid plant (from [19]).

The greatest value of somatic hybrids in citrus cultivar improvement is their use, when fertile, as tetraploid pollen parents in interploid crosses to generate seedless triploids [25, 40]. Seedlessness is one of the most important characteristics for citrus on the fresh fruit market. Due to several anomalies during sporogenesis [41] citrus triploids are generally sterile, therefore the creation of triploid hybrids is an important breeding strategy to develop new commercial citrus varieties. An effective way to recover triploid citrus plants is by hybridization between diploid and tetraploid parents [41, 42]. The availability of fertile and optimal quality tetraploid breeding parents is, therefore, crucial in citrus variety improvement.

Somatic hybridization by protoplast fusion is a very powerful tool in order to obtain an unlimited number of tetraploid parents for use in interploid hybridization. We have obtained several interspecific and intergeneric tetraploid somatic hybrid genotypes that have been used as pollen parent in interploid 2nx4n sexual hybridizations breeding programs aimed to recover citrus triploid progenies [43]. The recovered triploid progenies are now under selection for several breeding objectives, beyond seedlessness. For acid lemon improvement, primary breeding objectives include improved cold-hardiness, disease resistance and potential new industrial oils. With respect to the last topic, we have used specific allotetraploid somatic breeding parents that have been used as male parent in sexual back-crosses with diploid "Femminello" lemon in an effort to select triploid seedless lemon types showing improved traits with regard to essential oil quality and characteristics [44]. A large population of triploid hybrids has been obtained [43].

Asymmetric somatic hybridization and cybridization are considered as very promising approaches in crop breeding programs. Asymmetric somatic hybridization is a technique based on the somatic fusion between irradiated "donor" protoplasts (in which partial or total chromosome elimination has been carried out using rising doses of X or gamma rays, UV or restriction endonucleases) and "recipient" protoplasts, whose organelle genomes have been metabolically inhibited by iodoacetate (IOA). As a result, the heterokaryons combine vital cytoplasm from donor parent with wholesome nucleus from recipient parent (cybrids).

Dudits et al. [45] reported the first intergeneric asymmetric hybrid between parsley and tobacco protoplasts. Since them, many asymmetric somatic hybrids have been produced, and some of them show improved characteristics owing to the transfer of interesting traits from the donor species [46, 47].

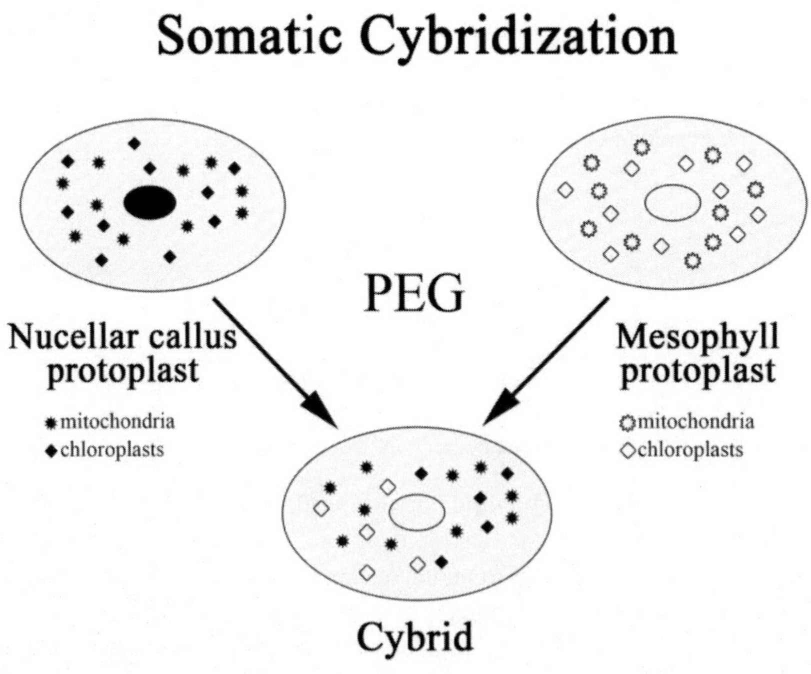

Figure 2. A schematic framework for the obtaining of *Citrus* cybrid plant.

The first citrus cybrid plants arose spontaneously, and unexpected, as consequence of symmetric somatic hybridization experiments in which diploid plants with morphology and zymotype of the presumably non-embryogenic leaf-mesophyll parent were recovered [48-52]. Cytological and molecular analysis revealed that the regenerated diploid plants were not directly regenerated from leaf-mesophyll protoplasts but rather were cybrids with the nuclear DNA from the leaf parent and the mtDNA from the corresponding embryogenic parent [23, 53-56]. Since then, citrus somatic cybrids have been regenerated from over 40 parental combinations by standard symmetric fusion [57], even if the regeneration mechanism is still not clear [58]. The importance of cybrids is due to a role of the organellar genome in plant growth and development. Although it is recognized that the nucleus possess the majority of genetic information it is, also, known that cytoplasm may alter the expression of nuclear genes through a communication termed retrograde signaling [59, 60]. In recent studies [61], comparing the proteomic profiles of cybrid plants and their corresponding fusion parent, a differentially expressed proteins was shown. The results suggested an alteration in photosynthesis, stress resistance and metabolism in the cybrid plant, providing useful information for assessing the resources and potential utilization of the cybrid and the importance of their nucleo-cytoplasmic arrangement related to fruit performance.

The mitochondrial DNA plays a key role in synthesis of the most relevant biomolecules (carbohydrates, lipid, amino-acids, vitamins and phytormones) and the generation of adenosine triphosphate (ATP). The production of cybrids is, therefore, of great interest in Citrus genetic and breeding. It has been, in fact, shown that cybridization can have an impact on aroma, fruit taste and nutritional quality [62, 63]. Fanciullino et al. [62] found that, for volatile compounds of leaves, although the cybrids were very close to their nuclear parent, some nucleo-cytoplasmic interactions occurred, such as the synthesis of more monoterpene alcohols and sesquiterpene in citrus cybrids [62]. It has been showed that cybridization in citrus affects organic acid content although no modification was observed for sugar and carotenoid content [63].

Compared to nuclear DNA, inheritance of cpDNA and mtDNA in somatic hybrids is relatively complex. Somatic fusion, in fact, can yield a combination of cytoplasms from different sources, unlike sexual cross hybridization that leads to maternal inheritance of cytoplasmic genomes. For Citrus fusion, the general consensus is that cpDNA is randomly transmitted; as for mtDNA, nearly all hybrids get theirs from the suspension parents, described cpDNA coexistence in Citrus fusions [64, 65].

Essential oil analysis in our diploid and tetraploid “Femminello” lemon cybrids confirmed the dominant role of the “Femminello” nuclear parent in determining the peel oil characteristics and the apparent marginality of the organellar parent “Valencia” sweet orange. However, the dominance of the “Femminello” nuclear parent is more marked in the 2n rather than in 4n cybrid [30]. Several characteristics (i.e., double amount of oxygenated monoterpenes in 2n cybrid as compared to 4n cybrid) make us think that chromosomal rearrangements may occur during cybrid formation as consequence of the heavy genetic manipulation during protopast fusion and regeneration process [30]. The presence of distinctive traits in the essential oil profiles is a further demonstration of the high potential of somatic hybridization or cybridization in citrus cultivar improvement.

B. Molecular Genetics

Genetic engineering and biotechnology hold great potential for plant breeding, expediting the time taken to produce crop varieties with desirable characters. With the use of molecular techniques it now possible to hasten the transfer of desirable genes among varieties and to introgress novel genes from (e.g.) related wild species. Mapping and sequencing of plant genomes help to elucidate gene function, gene regulation and their expression.

Molecular markers are used to identify and tag desired genes. Linkage can define the genetic distances between polymorphic traits which may be recognized as differences in appearance of enzyme activities, restriction fragment lengths or nucleotide sequences at an allelic locus. These polymorphisms are revealed by molecular markers such as Amplified Fragment Length Polymorphism (AFLP), Restriction Fragment Length Polymorphism (RFLP), microsatellite or Simple Sequence Repeat (SSR), Random Amplified Polymorphic DNA (RAPD), Cleavable Amplified Polymorphic Sequences (CAPS), Sequence Characterized Amplified Regions (SCAR), Inter Simple Sequence Repeat (ISSR), Sequence Related Amplified Polymorphism (SRAP) and Single Nucleotide Polymorphism (SNP).

Molecular markers which are stable, detectable in all tissues and independent of environmental conditions, were used in *Citrus* in a wide range of application: a) cultivars identification and phylogenetics [66-72]; b) construction of linkage map [73-76] for Marker-Assisted Selection (MAS) and map-based cloning of genes. Molecular markers are useful for identifying and selecting genotypes with favorable traits and markers linked to traits of interest can be used to select for a desirable trait in a segregant population and facilitate crop improvement through MAS.

More various types of molecular markers have been used to characterize Citrus varieties and germplasm accessions (AFLP, RFLP, RAPD, SCAR, SSR, ISSR, SNP) [68-72, 75-80]. The increase of available DNA sequences, particularly expressed sequence tags (ESTs) has provided new opportunities for development of molecular markers for *Citrus* sp. [81].

Microsatellites have proven to be the marker of choice in different plant species and *Citrus* breeding research, because of their variability, ease of use, accessibility of detection and reproducibility [73, 82-86]. The SSR are abundant and distributed throughout the genomes of eukaryotes. They are co-dominant markers, usually located in noncoding DNA regions, which can accumulate mutations more easily than coding regions and widely utilized in molecular genetic studies for mapping, fingerprinting, genetic diversity, and phylogenetic reconstruction. The microsatellite are short sequence elements composed of tandem repeat units, characterized by a 1- to 6 base pairs (bp) in length [87]. The tandem repeats are ubiquitous and can be found in nuclear, chloroplast, and mitochondrial genomes. One of the main features of SSR is that the level of polymorphism can be highly discriminating, sufficient enough to show unique, specific profiles for each individual in a population from relatively few markers [88]. The elevated level of polymorphism is the consequence of the high mutation rates of these sequences, which allow variability in species otherwise characterized by low levels of genetic diversity [89]. Therefore, microsatellite markers are the most commonly utilized markers in molecular biology for mapping, genetic diversity, phylogenetic construction, and fingerprinting.

Most of the research with molecular markers in *Citrus* was based on the nuclear genome. However, the conservative nature of the chloroplast genome makes it more useful for parentage and taxonomic studies. The chloroplast genome is non-recombinant, uni-parentally inherited, haploid, small and the gene order is relatively conserved [77, 79, 90-92], properties that overcome the disadvantages encountered in analyses of nuclear DNA markers and useful to identify the nature of plastidial genome in the genotypes obtained by somatic protoplast fusion. Chloroplast microsatellite markers (cpSSR) were successfully used in studies on more species such as barley [93], rice [94], soybean [91], as well as higher woody angiosperm plants: *Olea europaea* L. [95], *Corylus avellana* L. [96], *Fagus sylvatica* L. [97]. Since in *Citrus* understanding taxonomy, phylogenetic relationships, and genetic variability are critical for determining genetic relationships, characterizing germplasm, designing sampling strategies, establishing breeding programs and for the isolation/registration of new cultivars with specific behavior and/or features, cpSSRs were developed [79, 98] and widely used [99].

Since the cytoplasm in *Citrus* is maternally inherited [100, 101], but in some hybrids an apparent biparental inheritance of mitochondria was observed [102], it is also important to study and classify the mitochondrial genome (mtDNA).

More polymorphism was found within the chloroplast genomes than within the mitochondrial genomes of the *Citrus* genus [103], however, even if the genetic role of mtDNA appears to be universally conserved, this genome shows remarkable variation in conformation and size [104]. The mtDNA region showed a low level of polymorphism, probably due to a slower nucleotide substitution rate in *Citrus* mtDNA [105], nevertheless a good set of mtSSRs were developed [106]. Starting from less conserved regions of the mitochondrial genomes, as intergenic sequences of nad3-rps12, rpl5-cob clusters or intronic regions of nad2, nad5, and nad7, a panel of polymorphic markers was isolated to study maternal mitochondrial inheritance. The selected mtDNA markers are very useful and efficient tools to determine the maternal origin of citrus species [106].

Although microsatellites are a flexible and widely used nevertheless require a substantial investment of time and money to isolate and plants with developed SSR markers are still very limited, especially for orphan crops and long-juvenile woody species [107]. In addition, developing techniques to support crops improvement, e.g., Genome-Wide Association Studies (GWAS) a powerful approach for mining novel functional variants [108, 109], needs high performance markers and systems that simultaneously analyzed a very large number of accessions for genetic variation underlying diverse complex traits.

The development of high-throughput NGS (Next Generation Sequencing) technologies and the possibility to sequence and re-sequence entire genomes or entire transcriptomes more efficiently and economically than the first generation sequencing strategies, allowed to obtain large-scale SNP identification per species and the onset of efficiency SNP genotyping platform [110]. Moreover, the high-throughput multiplexed SNP assay is a useful tool to perform genome-wide evaluation of allelic variation, the construction of high-density maps together with quantitative trait loci (QTL) mapping, the evaluation of genetic diversity and population structure and parentage analysis in crop plants. Examples included a number of plants both with moderate genome size (*Arabidopsis* and rice) [111, 112] and with large and highly complex genome, as soybean [113], tomato [114] and wheat [115], together with woody crop plant, such as apple [116], peach [117] and *Citrus* [118].

During the long history of evolution and cultivation of *Citrus* sp. has been developed many cultivars contain a variety of secondary metabolites, such as flavonoids [119], carotenoids [120] and limonoids [121, 122], which have health-promoting functions in humans. Therefore, it has become important to breed new cultivars enriched with such substances, increasing e.g., the content of β-cryptoxanthin, a carotenoid component with cancer preventative activity [120]. Quantitative trait loci (QTL) analysis is a powerful approach to map the genetic factors for important traits, such as β-cryptoxanthin content [123]. Hence, the development of SNPs is becoming necessary and high-throughput technologies are required to perform genotyping with a large number of markers and progenies.

SNPs are the most frequent type of variation found in the genome. As EST databases and whole genome sequences grow in availability, SNPs have become the most abundant and powerful polymorphic codominant markers [124]. Therefore, SNPs allow the implementation of very dense genetic linkage maps in plants [125]. Moreover, SNPs are generally considered to have a high identity by descent rate, and thus, they are very useful for genetic association studies [72, 126].

In addition to SNPs identification, it also become important to develop technologies to analyze many samples simultaneously, genotyped with thousands of SNP. The GoldenGate® Assay (Illumina Inc.) a SNP genotyping technologies, was applied successfully to various plant species, such as rice (*Oryza sativa* L.) [127], barley [128], soybean [129], white and black spruce [130], loblolly pine [131], sugi [132] and vitis [133]. Because the assay usually is capable of multiplexing from 96 to 1,536 SNPs in a single reaction, the high-throughput genotyping system can quickly generate high-density genetic maps for MAS of various important traits. A prototype 384 SNP *Citrus* array, named CitSGA-1, for the GoldenGate® Assay, was developed [120]. The number of chromosomes in haploid citrus is generally nine and *Citrus* SNP genotyping throughput covers all chromosomes, with an average of 43 SNPs/chromosome. Therefore, the prototype 384 SNP array is useful to map the breeding trait loci at the marker density depending on the number of heterozygous loci for each cultivar [120]. However, the information obtained by the developed prototype would allow to construct a higher multiplex custom assay made by over 1,500 SNPs, a new important system to identify SNP associated to traits of interest such as genes involved in secondary metabolites production.

CHEMISTRY OF *CITRUS* FRUITS

Citrus fruits are one of the richest natural source of nutrient components. Carbohydrate such as fructose, glucose and sucrose, together with non starch polysaccharides, pectin, cellulose and hemicellulose are present in large amount. The fat levels are very low, whereas the content of potassium is very high with respect of that of sodium, citrus fruits are also well known for their high content of vitamin C. Besides these components which possess a great nutritional impact, all citrus fruits synthetize and accumulate several class of phytochemical components, well known as secondary metabolites or natural products:

polyphenols (flavonoids, organic acids, anthocyanins, polymethoxy flavones): flavanones are normally the most represented flavonoids, flavones and flavonols are present in lower amount; anthocyanidins have been detected only in blood orange varieties, being responsible for their pigmentation; polymethoxy flavones are present almost exclusively in the *Citrus* genus, being concentrated in the peel and for this reason represent a portion of the peel *Citrus* essential oils, which are obtained mechanically;

coumarins: these components can be considered cinnamic acid-derived phenolic compounds with fused benzene and α-pyrone rings. They have been mainly found in the Rutaceae and Umbelliferae families and possess appetite-suppressing properties, then showing a defensive function discouraging animals from eating plants containing them;

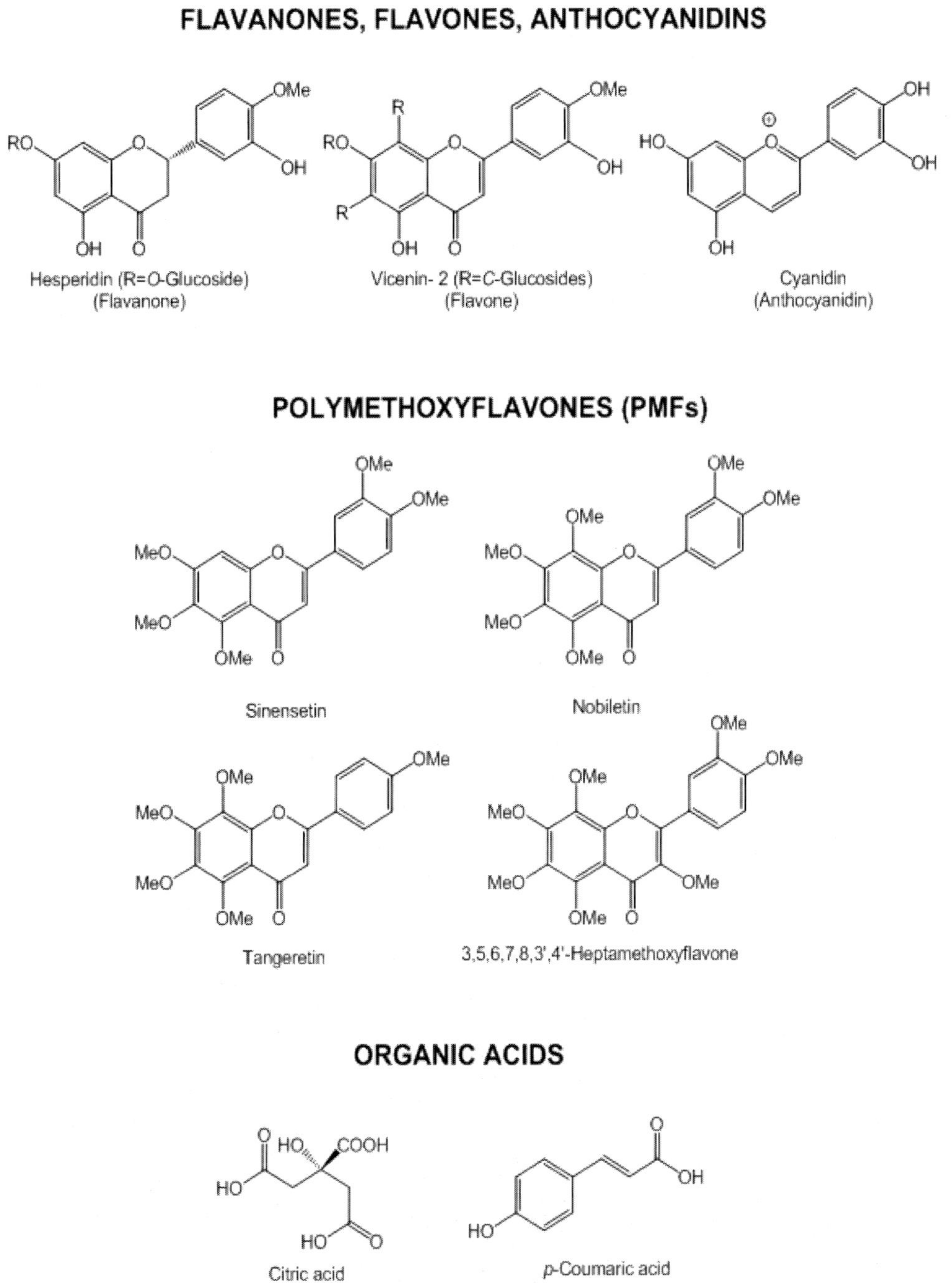

Figure 3A. A first selection of secondary metabolites from *Citrus* fruits.

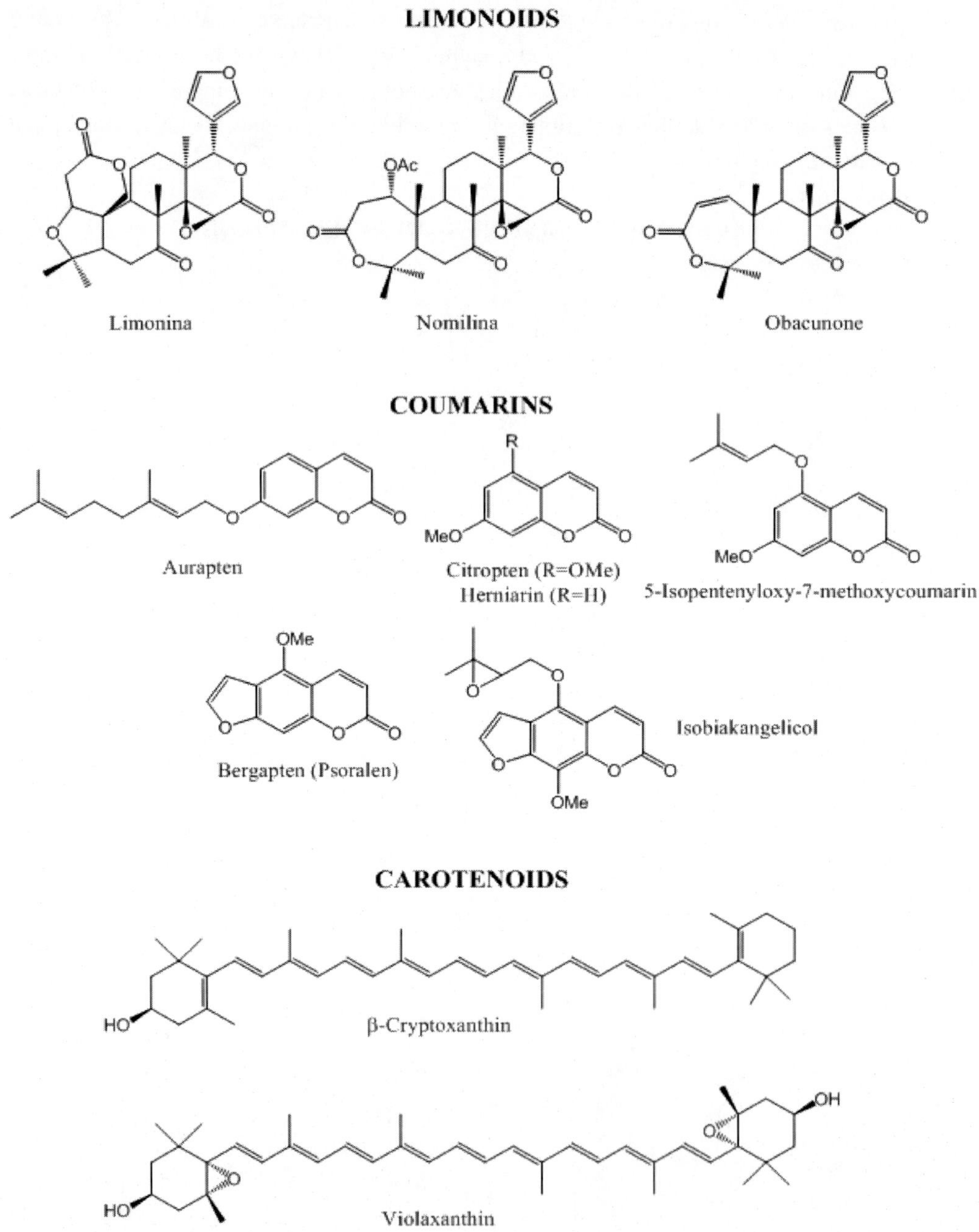

Figure 3B. A second selection of secondary metabolites from *Citrus* fruits.

limonoids: these compounds known also as tetranorterpenoids are present mainly present in seeds and the albedo portion of the peel *Citrus* fruits. They are responsible for the bitterness of the fruits, and as previously mentioned for coumarins, limonoids possess antifeedant properties [134];

carotenoids: several carotenoids have been found mainly in the flavedo portion of the peel and in the juice sacs, however the main components are violaxanthin (all-*trans* and 9-*cis* isomers) and β-criptoxanthin [135, 136];

terpenoids: besides limonoids and carotenoids previously described, most of the terpenoids derivatives found in *Citrus* fruits are present in their essential oils as mono and sesquiterpenes in the hydrocarbon and oxygenated form [137, 138].

The molecular structure of a selected number of these components is reported in the Figures 3A and 3B.

Another significant class of phytochemical components, as previously mentioned, is represented by peel essential oils [137, 138]. The volatile components found in the *Citrus* peel essential oils largely belong to the terpene class of compounds, being mono and sesquiterpene hydrocarbons the main components together with lower amount of the corresponding oxygenated derivatives (Figure 4).

Most of the aforesaid compounds are bioactive products which have been tested for many biological activities, but they in these last years are successful as nutraceutic, namely as products with a strong and efficacious preventive action against several pathologies, which can be assumed with a diet normally rich in fruits and vegetables, as the Mediterranean diet [139-142].

Another important and not marginal aspect of the study of the secondary metabolite profile of a particular plant is that this study represents an extraordinary and, for many aspects, irreplaceable tool for establishing traceability, originality and typicalness of a given plant material, as well as of its processing products. This is today testified by many studies on a large number of products coming from the different biomolecular pathways operating in the plant material, the polyphenol metabolic pool, almost ubiquitous in the plant kingdom is a real example of this trend [143]. Taxonomic implications, geographic discrimination and biodiversity defense represents some of the milestones of these studies [144-148].

NEW HYBRIDS

A. General

In these last years we have analysed several new *Citrus* hybrids obtained with different crossbreeding procedures with aim to characterize new varieties with an hopeful increased amount of nutraceutical components, and/or new and original aromatic profiles due to a different combination of volatile components [30, 44, 149-156]. Both these aspects should be useful for the introduction of these new fruits into the fresh market and in the industrial chain of transformation.

An hybridization project possesses a multi-purpose aim accounting for the production of seedless fruits with an optimal size, easily removable peel, new and original organoleptic features, and possibly with precocious or late ripening, moreover, not secondary aspects lie in the possibility to obtain plants with improved resistance against biotic and abiotic stresses [157, 158].

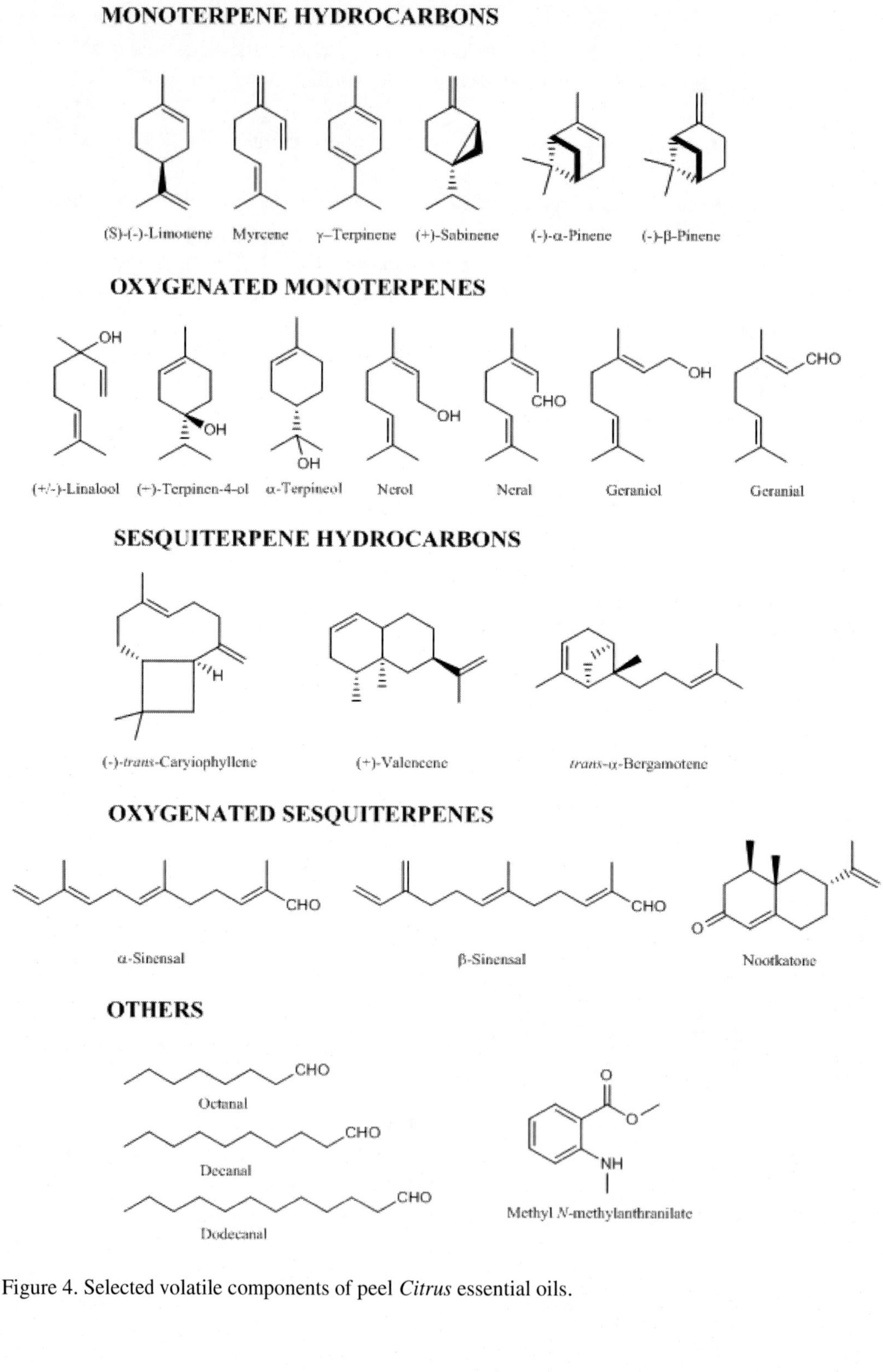

Figure 4. Selected volatile components of peel *Citrus* essential oils.

B. Hybrids from "Valencia" Sweet Orange and "Femminello" Lemon. Somatic and Sexual Hybrids

The first example here reported regarding the study of the chemical profiles of new *Citrus* hybrids concerns the hybrids from the crossbreeding of the "Valencia" sweet orange (*Citrus sinensis* L. Osbeck) + the "Femminello" lemon (*Citrus limon* L. Burm) to obtain the allotetraploid somatic hybrid (Valencia + Femminello), which in turn by back crosses with "Femminello" lemon allowed to obtain three sexual hybrids, named Fx(V + F)5, Fx(V + F)7 and Fx(V + F)10.

The somatic hybrid V + F is quite similar in shape, size and colour to that of "Valencia" sweet orange parent, whereas fruit aspect of the three selected sexual hybrids, Fx(V + F)5, Fx(V + F)7 and Fx(V + F)10, coming from interploid (4n x 2n) backcross with "Femminello" lemon was similar to that of the *C. limon* maternal parent, with the partial exception of Fx(V + F)7, which shows a color and a size more similar to the V + F parent [155].

The polyphenol profile of the juices of all the aforesaid samples, namely parents and hybrids, has been studied by a combination of liquid-chromatography-ultraviolet-diode-array-detector-mass-spectrometry (LC-UV-DAD-MS). Ten flavonoids (TF), comprising seven flavanones and three flavones, and four hydroxycinnamic acids (HCA) have been characterized and quantified in all hybrids and compared with those of respective parents. The "Valencia" + "Femminello" somatic hybrid shows an intermediate polyphenol composition with respect to those of parents, with a slight prevalence of lemon influence. The three sexual hybrids show, instead, different and more complex chromatographic profiles.

The same aforesaid *Citrus* samples have been subjected to a further phytochemical study concerning the analysis of their peel essential oils [44]. The oils were obtained by hydrodistillation of the flavedo portion of the peel of each variety, and their composition has been studied by gas chromatography (GC) combined with a flame ionization detector (FID) and a mass spectrometry (MS). In total, 83 components were fully characterized and grouped in four classes (monoterpene hydrocarbons, oxygenated monoterpenes, sesquiterpenes, and others) for an easier comparison of all oils. The volatile profile of all hybrids is strongly influenced by that of "Femminello" lemon parent, however, a different behavior in the peel essential oil composition of the allotetraploid hybrid and the three hybrids is observed.

POLYPHENOL PROFILE. Table 1 lists the quantitative profile of the ten flavonoids (TF) and four hydroxycinnamic acids (HCA) reported in Figure 5, individuated in "Valencia" orange, "Femminello" lemon and their hybrids.

The profile of "Valencia" sweet orange (V) is characterised by the predominance of two flavanones, hesperidin and narirutin, which together constitute 93% of this subclass, whereas vicenin-2 is the main and almost unique flavone. Amongst HCA, the ferulic is the main one, sinapic and coumaric follow at after a considerable gap, caffeic is found at trace level. "Femminello" lemon shows a greater amount of TF than "Valencia" orange, hesperidin and eriocitrin being the most important flavanones (97% of total subclass); diosmetin 6,8-di-*C*-glucoside is the main flavone, followed by vicenin-2 and diosmin. Concerning HCA, also in this case ferulic is the main acid, unlike "Valencia" orange, sinapic and coumaric are present in higher amounts, caffeic is still at trace level. The composition of both parents does not present significant differences with the average composition of this species as reported in literature [159-164].

FLAVANONES

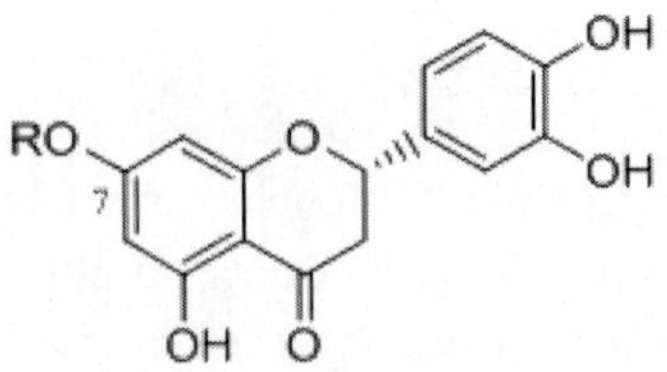

Eriocitrin R=7-*O*-rutinoside

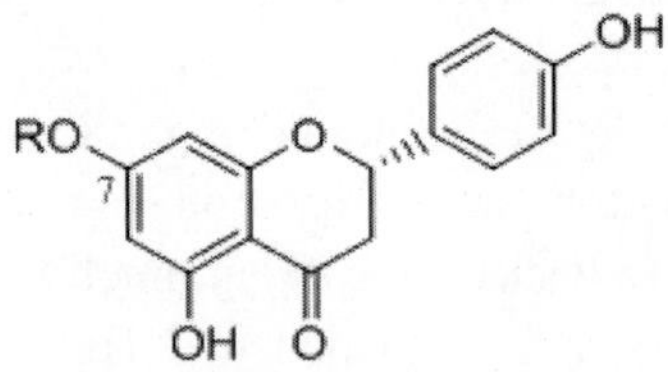

Naringenin R=H
Naringenin 7-*O*-glucoside R=7-*O*-glucoside
Narirutin R=7-*O*-rutinoside
Naringin R=7-*O*-neohesperidoside

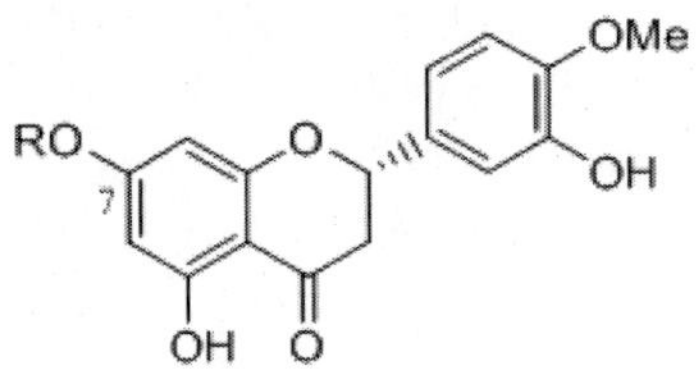

Hesperidin R=7-*O*-rutinoside
Neohesperidin R=7-*O*-neohesperidoside

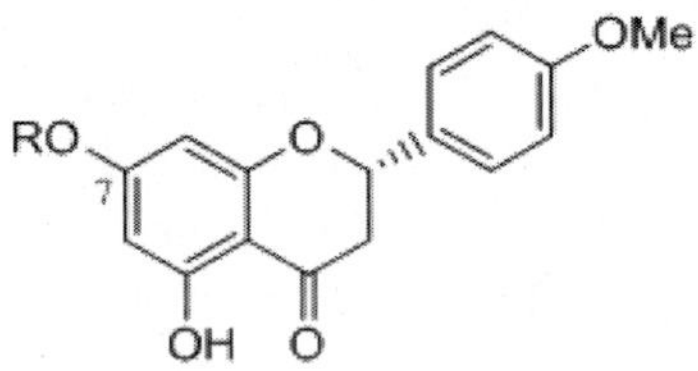

Didymin R=7-*O*-rutinoside

FLAVONES

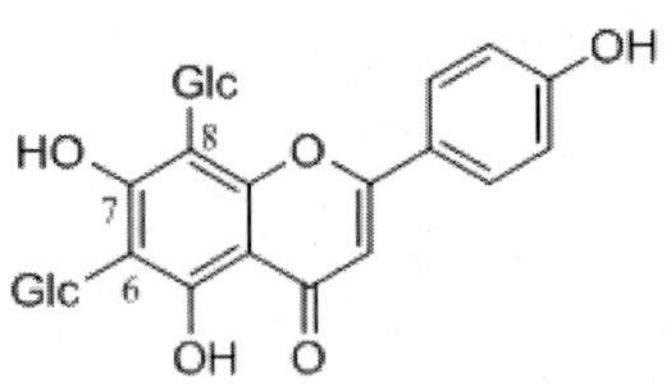

Vicenin-2 (Glc=*C*-glucoside)

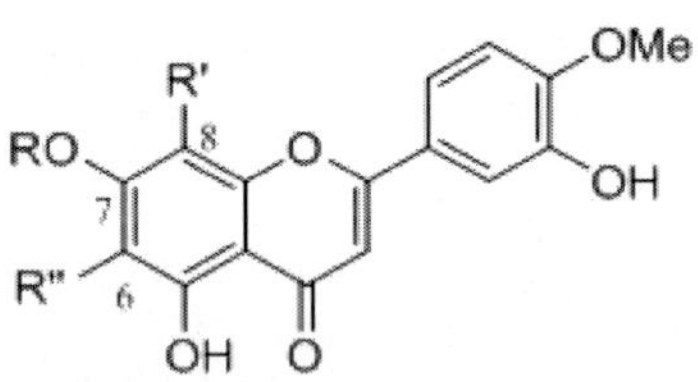

Diosmetin 6,8-di-*C*-glucoside R=H, R'=R"=*C*-glucoside
Diosmin R= 7-*O*-rutinoside, R'=R"=H

HYDROXYCINNAMIC ACIDS

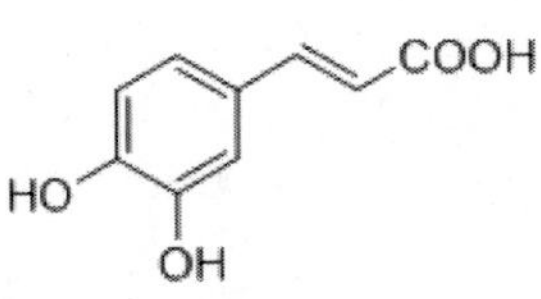

Caffeic acid

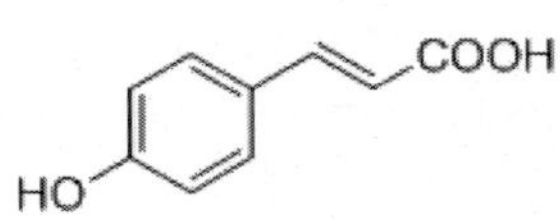

Coumaric acid

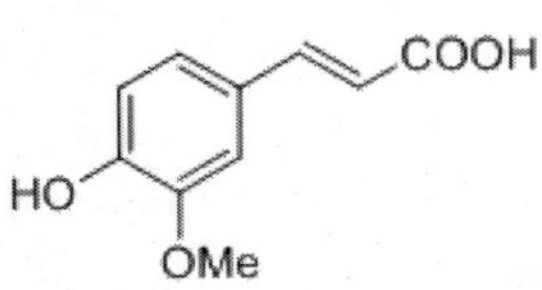

Ferulic acid

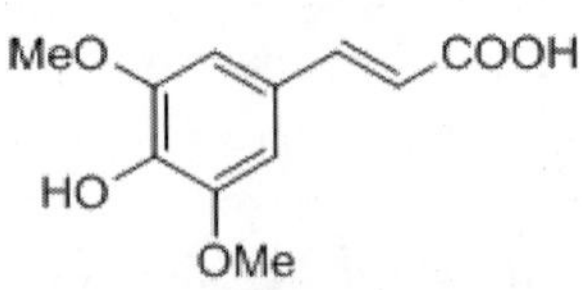

Sinapic acid

Figure 5. Polyphenols from *Citrus* juices.

Table 1. Polyphenols (mg/L) in juices of "Valencia" orange, "Femminello" lemon and their hybrids

#	Components	Valencia (V)	Femminello (F)	Valencia + Femminello (V + F)	F x (V + F)5	F x (V + F)7	F x (V + F)10
	Total Flavonoids	**135.64**	**346.42**	**243.13**	**167.33**	**267.90**	**381.18**
	Flavanones	**108.04**	**199.84**	**182.17**	**79.80**	**235.26**	**261.79**
3	Eriocitrin	0.65 (0.08)^{a}	70.74 (2.53)^{d}	54.22 (0.05)^{c}	43.45 (0.68)^{b}	82.46 (0.63)^{e}	112.12 (1.39)^{f}
4	Narirutin	19.39 (1.10)^{e}	0.96 (0.26)^{a}	17.48 (0.27)^{d}	0.57 (0.33)^{a}	9.12 (0.17)^{c}	3.20 (0.27)^{b}
5	Naringenin 7-*O*-glc	1.04 (0.31)^{bc}	2.69 (0.27)^{d}	1.52 (0.35)^{c}	0.38 (0.13)^{a}	0.99 (0.06)^{bc}	0.64 (0.13)^{ab}
6	Naringin	1.82 (0.20)^{b}	0.08 (0.04)^{a}	1.30 (0.47)^{b}	0.02 (0.01)^{a}	0.13 (0.01)^{a}	0.12 (0.09)^{a}
7	Hesperidin	81.50 (0.22)^{b}	124.58 (25.91)^{c}	107.35 (0.43)^{bc}	35.09 (0.48)^{a}	141.20 (0.31)^{c}	144.89 (7.16)^{c}
9	Neohesperidin	0.64 (0.46)$^{n.s}$	0.73 (0.20)$^{n.s.}$	0.41 0.27)$^{n.s.}$	0.27 (0.01)$^{n.s.}$	0.76 (0.02)$^{n.s.}$	0.75 (0.02)$^{n.s.}$
10	Dydimin	2.99 (0.00)^{d}	0.06 (0.01)^{ab}	0.10 (0.00)^{b}	0.03 (0.00)^{a}	0.59 (0.02)^{c}	0.07 (0.05)^{ab}
	Flavones	**27.28**	**145.93**	**60.74**	**87.08**	**32.01**	**118.37**
1	Vicenin-2	26.50 (2.17)^{d}	13.46 (0.10)^{b}	25.21 (1.68)^{d}	10.27 (0.03)^{ab}	8.54 (1.60)^{a}	19.13 ((0.06)^{c}
2	Diosmetin 6,8-di-*C*-glc	0.41 (0.17)^{a}	109.82 (0.39)^{f}	35.21 (1.25)^{c}	48.53 (0.02)^{d}	17.19 (0.39)^{b}	57.89 (0.17)^{e}
8	Diosmin	0.37 (0.30)^{a}	22.64 (19.60)^{abc}	0.31 (0.16)^{a}	28.27 (0.15)^{bc}	6.28 (0.30)^{ab}	41.35 (8.48)^{c}
	Hydroxycinnamic Acids	**16.50**	**11.45**	**16.38**	**10.21**	**17.35**	**8.44**
1*	Caffeic acid	0.78 (0.00)^{d}	0.27 (0.01)^{a}	0.68 (0.00)^{c}	0.46 (0.04)^{b}	1.05 (0.00)^{e}	0.30 (0.01)^{a}
2*	Coumaric acid	1.17 (0.00)^{a}	2.33 (0.03)^{c}	0.98 (0.00)^{a}	1.90 ((0.31)^{b}	1.97 (0.05)^{b}	0.97 (0.13)^{a}
3*	Ferulic acid	13.05 (0.04)^{e}	6.59 (0.04)^{c}	11.72 (0.04)^{d}	5.31 (0.42)^{b}	12.94 (0.01)^{e}	3.84 (0.07)^{a}
4*	Sinapic acid	1.51 (0.02)^{ab}	2.26 (0.08)^{c}	1.74 (0.02)^{b}	2.58 (0.22)^{d}	1.39 (0.00)^{a}	2.67 (0.04)^{d}

Values represent averages of three determinations and standard deviation (± SD) is given in parentheses; different letter in the same row represents significant difference at $p \leq 0.01$ by Duncan's multiple range test; n.s. = non significant.

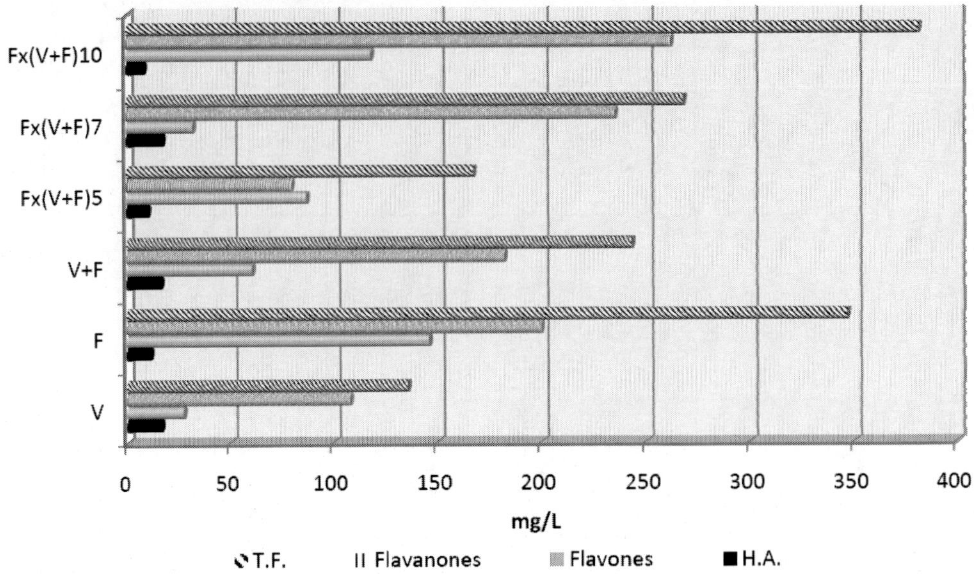

Figure 6. Content (mg/L) of total flavonoids (T.F.), comprising flavanones and flavones, and hydroxycinnamic acids (H.A.), in the juices of "Valencia" orange (V), "Femminello" lemon (F) and their hybrids: V + F, Fx(V + F)5, Fx(V + F)7 and Fx(V + F)10.

The compositional polyphenol profile of the allotetraploid somatic hybrid V + F is intermediate between those of parents, with a slight prevalence of lemon influence. Hesperidin and eriocitrin are the main flavanones as in lemon, analogously diosmetin 6,8-*C*-glucoside, as in lemon, is the most important flavone. Concerning the HCA, a closer behaviour to "Valencia" orange is instead observed.

The sexual hybrids show a variegated profile. In fact, the hybrid Fx(V + F)5 is the sole sample of this group showing a slight predominance of flavones with respect to flavanones, and unlike both parents eriocitrin is predominant with respect to hesperidin. The hybrids Fx(V + F)7 and Fx(V + F)10 show a higher level of flavanones than both parents, with the usual predominance of hesperidin over eriocitrin; however, unlike the previous hybrid, a higher amount of naritutin is observed. Instead the flavones content is quite different: Fx(V + F)7 has a lower concentration of these components than both parents, whereas Fx(V + F)10 has an intermediate amount, with the exception of diosmin, whose amount is decidedly higher than both parents. The HCA amount of Fx(V + F)5 and Fx(V + F)10 are comparable to lemon, whereas those of Fx(V + F)7 are much more similar to those of V + F. The graph of Figure 6 gives an immediate comparison of the content of the previous described components in orange, lemon and their hybrids.

In order to refine these quantitative data with the aim to obtain an optimal differentiation of all species here involved, namely parents and hybrids, a multivariate analysis using the quantitative data of total flavonoids (TF) and hydroxycinnamic acids (HA) has been applied. The discriminant analysis afforded the separation shown in Figure 7. The graphic representation in the two functions: namely, functions 1 (94.1%) and 2 (3.7%) representing 97.8% of the total variance, shows the large differentiation between "Valencia" orange and "Femminello" lemon. The somatic hybrid V + F is placed between both parents according to the two functions, being closer to the Femminello parent in both cases. The three sexual hybrids, Fx(V + F)5, Fx(V + F)7 and Fx(V + F)10, show a more irregular behaviour, confirming the previous considerations. It should be underlined that in this case their parents, namely F and V + F, are closer than the previous two parents V and F, in particular for the first and more significant function.

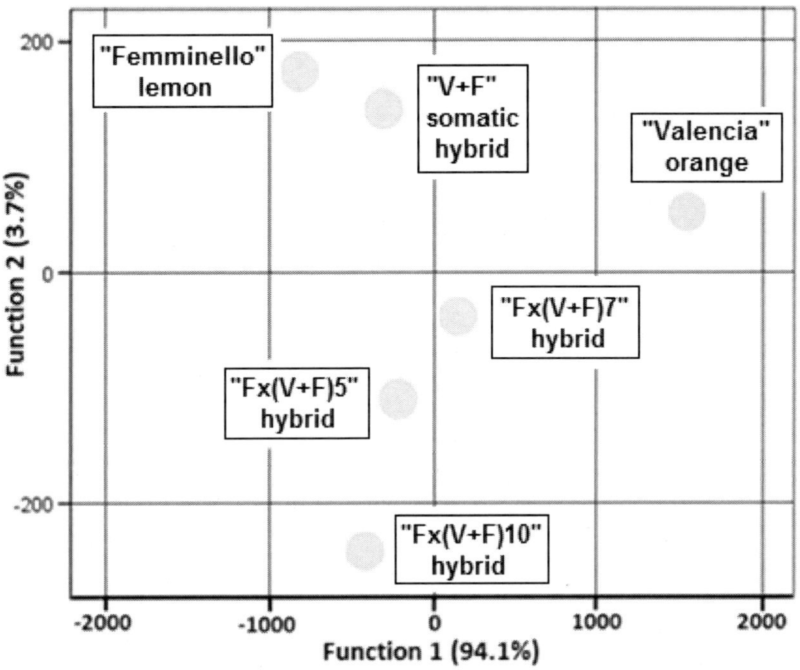

Figure 7. Discriminant score plot (Functions 1 and 2) of polyphenols (TF and HA) in the juices of "Femminello" lemon, "Valencia" orange, and their hybrids V + F, Fx(V + F)5, Fx(V + F)7 and Fx(V + F)10.

ESSENTIAL OIL PROFILE. The chemical composition of the essential oils of the hybrids and their parents, obtained by the peel hydrodistillation, have been studied by gas chromatography (GC) combined with a flame ionization detector (FID) and a mass spectrometry (MS), and Table 2 lists the composition of the essential oils of all samples. The 83 identified components were grouped into four classes: monoterpene hydrocarbons (15 components), oxygenated monoterpenes (31 compounds), sesquiterpenes (20 compounds) and others (17 compounds). Monoterpenes, both hydrocarbons and oxygenated, were the most highly represented classes: the former with a range of 76-97% and the latter with a range of 2-20%. The sesquiterpene and other classes were in all cases the least represented.

Table 2. Essential oil chemical composition of "Valencia" orange, "Femminello" lemon and V + F, F(V + F)5, F(V + F)7, F(V + F)10 hybrids

#[a]	Compounds	"Valencia" orange (V)	"Femminello" lemon (F)	(V + F)	F x (V + F)5	F x (V + F)7	F x (V + F)10
		%					
1	*n*-Octane[b]	0.02	0.06	0.01	0.02		0.02
2	*trans*-2-Hexenal[b]	0.02					
3	α-Thujene[b]		0.22	0.20	0.26	0.23	0.30
4	α-Pinene[b]	0.38	1.01	0.98	1.27	1.18	1.38
5	Camphene[b]		0.05	0.02	0.05	0.03	0.06
6	Sabinene[b]	0.35	0.71	1.56	0.85	0.70	0.95
7	β-Pinene[b]	0.06	7.29	3.15	8.33	6.07	9.05
8	6-methyl-5-Hepten-2-one		0.05		0.02	0.02	0.02
9	Myrcene	1.51	1.12	1.64	1.32	1.44	1.32
10	Octanal[b]	1.42	0.20	0.07	0.25	0.48	0.24
11	α-Phellandrene			0.02		0.04	
12	*p*-Mentha-1(7),8-diene	0.07					
13	α-Terpinene[b]	0.02	0.28	0.18	0.23	0.17	0.24
14	Limonene	91.51	59.75	84.56	67.37	74.48)	64.57
15	*cis*-β-Ocimene	0.07	0.07	0.07	0.10	0.05	0.10
16	*trans*-β- Ocimene				0.09	0.03	0.11
17	γ-Terpinene[b]	0.10	5.65	4.65 (0.061)	7.14	5.75	7.87
18	*trans*-Sabinene hydrate[b]						0.02
19	*n*-Octanol[b]	0.26	0.02		0.05	0.11	0.05
20	α-Terpinolene[b]	0.03	0.38	0.27	0.36	0.31	0.39
21	Linalool[b]	1.78	1.14	0.22	0.45	0.46	0.45
22	Nonanal[b]	0.07	0.31	0.07	0.21	0.08	0.15
23	*cis*-Thujone		0.08				
24	*trans*-Thujone		0.02		0.04		
25	*p*-1,3,8-Menthatriene	0.04					
26	Menth-2-en-1-ol			0.02			
27	*cis*-*p*-Mentha-2,8-dien-1-ol	0.02		0.02	0.02		
28	*trans*-Limonene oxide	0.05					
29	Camphor[b]		0.14	0.01	0.05	0.05	0.04
30	Citronellal[b]	0.10	0.35	0.11	0.15	0.70	0.12
31	*trans*-Pinene oxide				0.06	0.01	0.07

#[a]	Compounds	"Valencia" orange (V)	"Femminello" lemon (F)	(V + F)	F x (V + F)5	F x (V + F)7	F x (V + F)10
		%					
32	*cis*-Chrysanthenol		0.07		0.03	0.02	0.03
33	Borneol[b]		0.05		0.03	0.03	0.02
34	*n*-Nonanol[b]	0.02			0.04	0.02	0.04
35	Terpinen-4-ol[b]	0.11	0.82	0.32	0.46	0.32	0.43
36	*trans-p*-Mentha-1(7),8-dien-2-ol				0.08	0.05	0.09
37	α-Terpineol[b]	0.24	1.41	0.09	0.77	0.55	0.79
38	Estragole[b]		0.30	0.04	0.03	0.05	0.03
39	Decanal[b]	0.43	0.06	0.04	0.05	0.14	0.05
40	Octyl acetate[b]		0.01		0.01	0.02	
41	*trans*-Carveol	0.08	0.02				0.02
42	Citronellol			0.17	0.02		0.02
43	*cis-p*-Mentha-1(7),8-dien-2-ol		0.03				
44	Nerol[b]	0.10	2.39	0.20	0.82	0.75	0.84
45	Neral[b]	0.23	4.64	0.09	2.70	1.44	3.16
46	Carvone	0.05		0.01	0.93		
47	Geraniol[b]	0.05	1.60	0.04	0.90	0.30	0.97
48	*trans*-2-Decenal[b]	0.01					
49	Geranial[b]	0.24	6.44	0.14	3.60	1.86	4.26
50	Perillaldehyde[b]	0.05		0.01	0.02	0.04	0.03
51	Limonen-10-ol		0.09	0.18	0.02		
52	Thymol[b]	0.15		0.10	0.02	0.71	0.02
53	Undecanol[b]				0.02	0.02	
54	Carvacrol[b]	0.03	0.14				
55	Undec-10-en-1-al					0.02	0.02
56	Terpinyl acetate			0.02		0.02	
57	Citronellyl acetate		0.02	0.07	0.07	0.07	0.03
58	Neryl Acetate		0.55	0.22	0.49	0.18	0.60
59	Geranyl Acetate		0.27	0.10	0.41	0.20	0.40
60	β-Elemene	0.02	0.04	0.01	0.02	0.02	
61	Decyl acetate[b]					0.02	
62	Dodecanal[b]	0.04		0.02			

Table 2. (Continued)

#a	*Compounds*	"Valencia" orange (V)	"Femminello" lemon (F)	(V + F)	F x (V + F)5	F x (V + F)7	F x (V + F)10
		%					
63	*N*-Methyl anthranilate		0.02				
64	α-*cis*-Bergamotene		0.11		0.03	0.01	0.02
65	Caryophyllene[b]	0.02	0.13	0.04	0.10	0.08	0.10
66	β-Copaene	0.01					
67	α-*trans*-Bergamotene		0.21	0.14	0.19	0.20	0.19
68	Humulene[b]		0.03	0.03			
69	β-*trans*-Farnesene						0.02
70	β-Santalene		0.01				
71	Germacrene D			0.01	0.02	0.02	0.01
72	Valencene	0.06	0.07	0.07	0.03	0.04	0.02
73	Biciclogermacrene		0.02		0.02	0.02	0.02
74	α-*cis*-Bisabolene			0.02	0.02		0.02
75	α-*E,E*-Farnesene		0.02	0.02		0.03	0.02
76	β-Bisabolene		0.40	0.21	0.31	0.32	0.30
77	δ-Cadinene	0.01		0.01			
78	*epi*-β-Santalene		0.03				
79	α-Bisabolol[b]		0.05	0.02	0.03	0.02	0.03
80	β-Sinensal	0.05		0.09			
81	α-Sinensal	0.02		0.02			
82	Notkatone	0.02	0.02				
83	Tricosane		0.03				
	Classes[c]						
	Monoterpene hydrocarbons	**94.15^{E}**	**76.53^{A}**	**97.28^{F}**	**87.39^{B}**	**90.49^{D}**	**89.06^{C}**
	Oxygenated monoterpenes	**3.25^{B}**	**20.30^{F}**	**2.14^{A}**	**12.13^{D}**	**7.75^{C}**	**12.41^{E}**
	Sesquiterpenes	**0.20^{A}**	**1.13^{E}**	**0.70^{B}**	**0.77^{D}**	**0.77^{D}**	**0.75^{C}**
	Others	**2.30^{F}**	**1.07^{E}**	**0.25^{A}**	**0.71^{C}**	**0.98^{D}**	**0.61^{B}**

[a] Compounds are listed according to the elution order, and values (relative peak area percent) represent averages of 3 determinations. [b] Co-elution with authentic sample. [c] Different letters in the same line represent significant differences at $p \leq 0.01$ by HSD Tukey test.

The "Femminello" essential oil is characterized by a lower amount of limonene and a higher content of β-pinene and γ-terpinene, if compared with other *Citrus* essential oil. Among the oxygenated monoterpenes, lemon presents an high content of neral and geranial (known as citral) and the corresponding alcohols nerol and geraniol [138]. The "Valencia" oil is, instead characterized by ca. 92% of limonene, as usually observed in an orange essential oil. The content of oxygenated monoterpenes is very low (ca. 3%), with linalool as the main component [138].

The essential oil profile of the allotetraploid somatic hybrid (V + F) presents a marked similarity with "Valencia" parent. However, the amount of the main components, namely the hydrocarbons limonene, γ-terpinene and β-pinene, in the new hybrid can be considered intermediate between those of both parents, whereas the content of oxygenated monoterpenes is lower than those of both parents (Table 2).

The aromatic profile of the three sexual hybrids: Fx(V + F)5, Fx(V + F)7 and Fx(V + F)10, shows an intermediate behavior between those of parents, even though a slight supremacy of "Femminello" lemon must be accounted. Limonene is again the main component and in all hybrids its content is between those of both parents, instead the amount of the other monoterpene hydrocarbons, namely α-pinene e γ-terpinene, in the hybrids 5 and 10 is higher than those measured in both parents, whereas in the hybrid 7 is decidedly lower. This different behavior of hybrid 7 is confirmed by the amount of other components, in fact, among the oxygenated monoterpenes the couple neral/geranial (citral), in all sexual hybrids (5, 7 and 10), sets at an intermediate level between those of parents, but closer to the lemon profile, however, in the hybrid 7 its content is again decidedly lower. This compositional feature is also confirmed by other minor components (Table 2), defining a general diversity of the essential oil profile of the hybrid 7 with respect to the other two sexual hybrids, 5 and 10. Therefore, the hybrids 5 and 10 possess a marked lemon character, whereas in the hybrid 7 a greater influence of the somatic hybrid V + F comes out.

The comparison of the content of the main classes of the essential oil components, namely monoterpene hydrocarbons, oxygenated monoterpenes, sesquiterpenes and others, is reported in Figure 8.

It is quite clear from this graph the compositional difference between the two typologies of hybrids: the strong similarity with 'Valencia' parent of the allotetraploid somatic hybrid (V + F), and the marked lemon character of the three sexual hybrids.

Also in this case as carried out with the polyphenol content, in order to obtain a best differentiation of all fruits involved in this study, all components of each essential oil were investigated by means of multivariate analysis, applying the linear discriminant analysis (LDA), which has been successfully applied in previous occasions in the differentiation of citrus juices as well as peel and leaf citrus oils [165-168].

Figure 9 shows the graphic representation of the variables (all essential oil components) in the two functions, 1 and 2, which contributed 95.0% and 3.7% of the total variance, respectively. The figure shows the expected large differentiation between "Valencia" orange and "Femminello" lemon; the V + F somatic hybrid is placed between parents according to the two functions, being closer to the "Femminello" parent for the first function and closer to the "Valencia" parent for the second function.

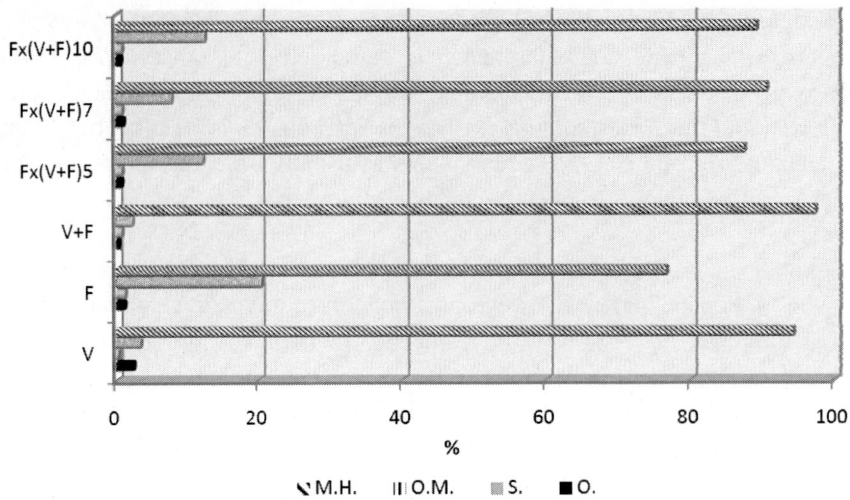

Figure 8. Content (%) of monoterpene hydrocarbons (M.H.), oxygenated monoterpenes (O.M.), sesquiterpenes (S.) and others (O.) in the essential oil of "Valencia" orange (V), "Femminello" lemon (F) and their hybrids: V + F, Fx(V + F)5, Fx(V + F)7 and Fx(V + F)10.

The behavior of the three sexual hybrids appear very similar, placing between the "Femminello" lemon and V + F parents regarding the most important function 1 in a similar position, and confirming the contribution of both parents in the elaboration of their peel essential oils. However, the hybrids 5 and 10 share the same position confirming a reciprocal substantial similarity, being closer to "Femminello" parent than the hybrid 7, which as mentioned shows a higher similarity with V + F parent.

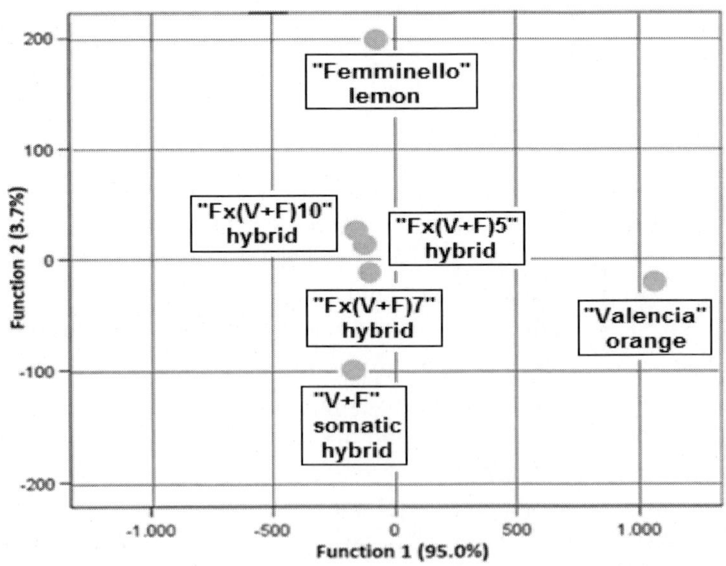

Figure 9. Discriminant score plot (Functions 1 and 2) of all components of the essential oils of "Femminello" lemon, "Valencia" orange, V + F allotetraploid somatic hybrid, and the sexual hybrids Fx(V + F)5, Fx(V + F)7 and Fx(V + F)10.

C. Cybrids from "Valencia" Orange and "Femminello" Lemon

Two new somatic hybrids, namely the 2n and 4n cybrids, have been obtained by protoplast fusion of "Valencia" sweet orange (*Citrus sinensis* L. Osbeck) + "Femminello" lemon (*C. limon* L. Burm). Their peel essential oils, as previously described, have been obtained by hydrodistillation and analyzed by a combination of GC-FID-MS.

Table 3 lists the composition of the essential oils of "Femminello" lemon, "Valencia" orange and their hybrids coming from the somatic fusion. In total, 87 components were fully identified covering more than 98% of the total composition. The components were grouped into four classes: monoterpene hydrocarbons (14 components), oxygenated monoterpenes (33 compounds), sesquiterpenes (23 compounds) and others (17 compounds), for an easier comparison of the oils [30]. As reported for the previously discussed *Citrus* varieties, monoterpenes, both hydrocarbons and oxygenated, were the most highly represented classes: the former with a range of 76-97% and the latter with a range of 2-20%. The sesquiterpene and other classes were in all cases the least represented (Figure 10).

The discussion of the chemical composition of "Femminello" lemon, "Valencia" sweet orange and their alloteraploid somatic hybrid V + F has been performed in the previous hybrid report. It is important to underline that the essential oil of the allotetraploid somatic hybrid (V + F) shows a more marked similarity with "Valencia" parent essential oil.

A different picture emerges from the analysis of the essential oil data of the two cybrids. In fact, in this case it is quite clear the dominant role, especially in the 2n rather than in 4n cybrid, of the "Femminello" lemon in the oil production of both cybrids, and, at the same time, the apparent marginality of the "Valencia" orange. Both hybrids present a slightly higher limonene content than lemon, whereas the other two main monoterpene hydrocarbons, β-pinene and γ-terpinene, are more markedly present in both hybrids. The 2n cybrid contain a double amount of oxygenated monoterpenes with respect to 4n cybrid, *ca.* 10 *vs. ca* 5%, mainly due to the different amount of the two couples of components (neral/geranial – nerol/geraniol) previously cited for the lemon parent. These four components, in fact, amount to *ca.* 7% in 2n cybrid, and *ca.* 3% in the 4n. The dominance of "Femminello" is also confirmed by the profile of the minor components, most of them below 1% (Table 3).

As in the previous cases in order to obtain a best differentiation of all fruits involved in this study, namely parents, 2n and 4n cybrids and allotetraploid V + F hybrid, all components of each essential oil were investigated by means of multivariate analysis, applying the linear discriminant analysis (LDA).

In Figure 11 the graphic representation of the variables (all essential oil components) in the two functions, 1 and 2, which contributed 79.3 and 15.5% of the variance, respectively, therefore, the combination of the first and the second function gives almost 95% of the total variance of the system. Figure 11 shows the well known large differentiation between 'Valencia' orange and 'Femminello' lemon, and the intermediate position between parents according to the two functions of the somatic hybrid 'V + F', being closer to the "Femminello" parent, as observed in a previous study dealing with the polyphenol profiles [155]. Both 2n and 4n cybrids, appear very similar, placing on an almost superimposable position with respect to the "Femminello" lemon parent regarding the most important function 1, clearly confirming the strong contribution of this parent in the elaboration of the peel essential oil of both cybrids.

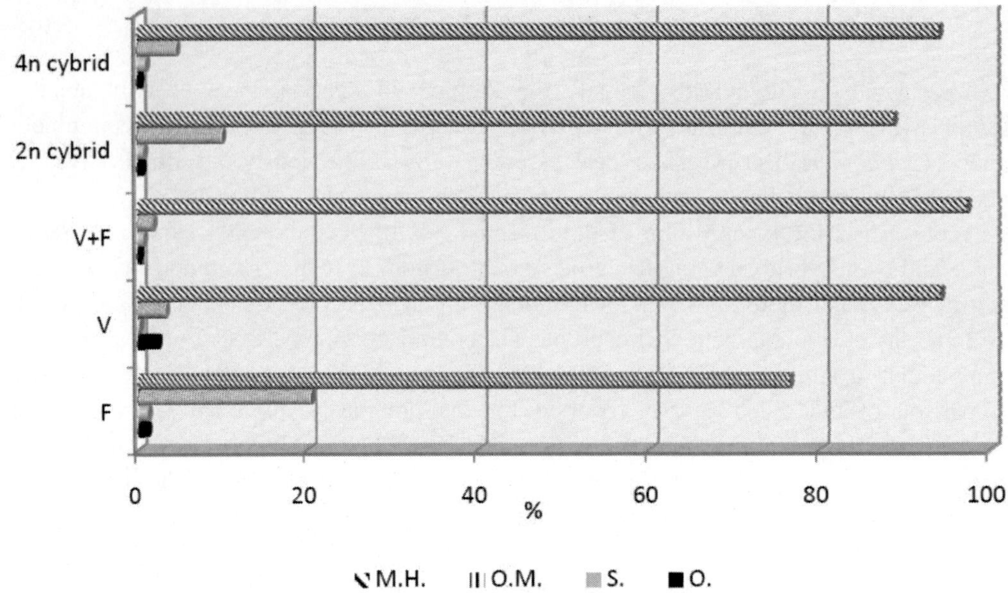

Figure 10. Content (%) of monoterpene hydrocarbons (M.H.), oxygenated monoterpenes (O.M.), sesquiterpenes (S.) and others (O.) in the essential oil of "Femminello" lemon (F), "Valencia" orange (V), V + F hybrid, 2n and 4n cybrids.

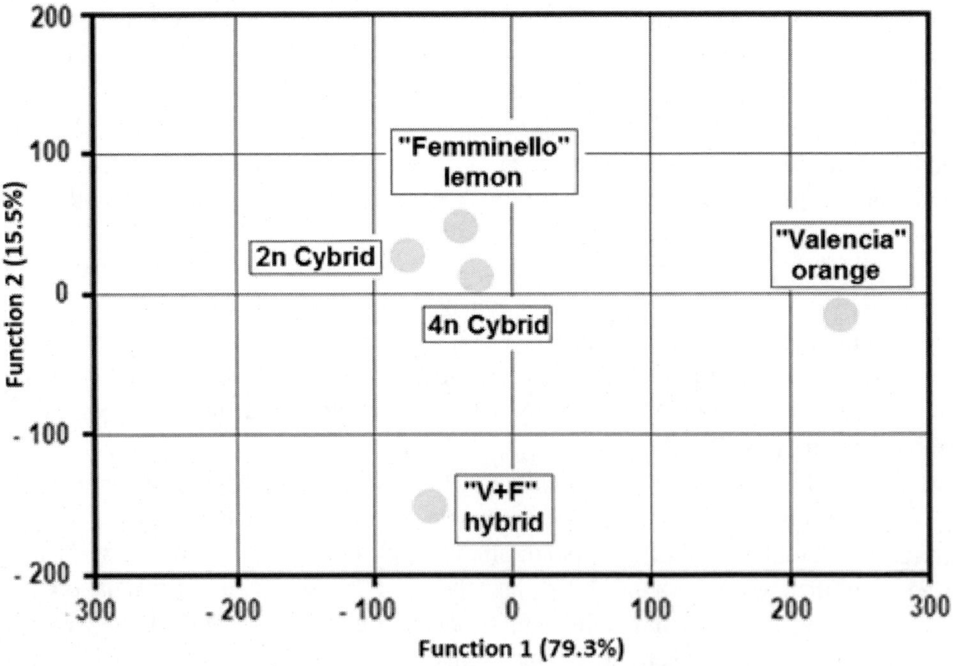

Figure 11. Discriminant score plot (Functions 1 and 2) of all components of the essential oils of "Femminello" lemon, "Valencia" orange, V + F allotetraploid hybrid, 2n and 4n cybrids.

Table 3. Essential oil chemical composition of 2n and 4n cybrids, V + F hybrid, “Femminello” lemon and “Valencia” orange[a]

Peak #	Compounds	“Femminello” lemon	“Valencia” orange	V + F hybrid	2n Cybrid	4n Cybrid
		%				
1	Octane[b]	t	t	0.02	0.03	0.01
2	Hexanal[b]	0.06	0.02			
3	*trans*-2-Hexenal		0.02			
4	α-Thujene[b]	0.22		0.20	0.33	0.32
5	α-Pinene[b]	1.01	0.38	0.98	1.51	1.50
6	Camphene[b]	0.05		0.02	0.06	0.06
7	Sabinene[b]	0.71	0.35	1.56	1.24	0.89
8	β-Pinene[b]	7.29	0.06	3.15	10.44	9.22
9	6-methyl-5-Hepten-2-one	0.05			0.02	
10	Myrcene[b]	1.12	1.51	1.64	1.33	1.44
11	Octanal[b]	0.20	1.42	0.07	0.19	0.05
12	α-Phellandrene	t	t	0.02	t	0.03
13	*p*-Mentha-1(7),8-diene		0.07			
14	α-Terpinene[b]	0.28	0.02	0.18	0.27	0.25
15	Limonene[b]	59.75	91.51	84.56	64.24	71.21
16	*cis*-β-Ocimene	t	t		0.02	
17	*trans*-β-Ocimene	0.07	0.07	0.07	0.14	0.16
18	γ-Terpinene[b]	5.65	0.10	4.65	8.64	8.49
19	*trans*-Linalool oxide (furanoid)				0.02	
20	*trans*-Sabinene hydrate[b]	t	t		t	
21	Octanol[b]	0.02	0.26		0.04	0.02
22	α-Terpinolene[b]	0.38	0.03	0.27	0.44	0.44
23	Linalool[b]	1.14	1.78	0.22	0.42	0.25
24	Nonanal[b]	0.31	0.07	0.07	0.20	0.10
25	*cis*-Thujone	0.08				0.01
26	*trans*-Thujone	0.02				
27	dehydro-Sabina Ketone			0.02		
28	*cis*-*p*-Menth-2-en-1-ol			0.02		
29	*cis-p-Mentha-2,8-dien-1-ol*		0.02	0.02	0.03	0.03

Table 3. (Continued)

Peak #	Compounds	"Femminello" lemon	"Valencia" orange	V + F hybrid	2n Cybrid	4n Cybrid
		%				
30	*trans*-Limonene oxide		0.05			
31	Camphor[b]	0.14		0.01	0.04	0.03
32	Citronellal[b]	0.35	0.10	0.11	0.10	0.07
33	β-Pinene oxide	t	t		t	0.02
34	*cis*-Chrysanthenol	0.07			0.04	
35	Borneol[b]	0.05			0.02	t
36	*n*-Nonanol[b]		0.02		0.05	0.03
37	*trans*-*p*-Mentha-1(7),8-dien-2-ol					0.03
38	Terpinen-4-ol[b]	0.82	0.11	0.32	0.50	0.35
39	α-Terpineol[b]	1.41	0.24	0.09	0.87	0.43
40	Estragole[b]	0.30		0.04	0.03	0.10
41	Decanal[b]	0.06	0.43	0.04	0.03	0.03
42	Octanol acetate[b]	0.01				
43	*trans*-Carveol	0.02	0.08		0.01	0.01
44	*cis*-*p*-Mentha-1(7),8-dien-2-ol	0,03				
45	Citronellol			0.17		
46	Nerol[b]	2.39	0.10	0.20	0.69	0.41
47	Carvone	t	0.05	0.01	t	
48	Neral[b]	4.64	0.22	0.09	2.46	0.75
49	Geraniol[b]	1.60	0.05	0.04	0.79	0.40
50	*trans*-2-Decenal[b]		0.01			
51	Geranial[b]	6.44	0.24	0.14	3.24	1.07
52	Perillaldehyde[b]		0.05	0.01	0.03	0.03
53	Limonen-10-ol		0.15		0.02	0.02
54	Thymol[d]	0.09		0.14	t	0.02
55	Carvacrol[b]	0.14	0.03			
56	Undecanal[b]					0.01
57	Undec-10-en-1-al	t	t		0.02	0.01
58	α-Terpinyl acetate			0.02		
59	Citronellyl acetate	0.02		0.07	0.03	0.05
60	Neryl acetate	0.55		0.22	0.38	0.40
61	Geranyl acetate	0.27		0.10	0.27	0.29
62	β-Elemene	0.04	0.02	0.01		
63	Dodecanal[b]		0.04	0.02		

Peak #	Compounds	"Femminello" lemon	"Valencia" orange	V + F hybrid	2n Cybrid	4n Cybrid
		%				
64	*N*-Methyl anthranilate	0.02				
65	*cis*-α–Bergamotene	0.11			0.01	0.02
66	β-Caryophyllene[b]	0.13	0.02	0.04	0.10	0.17
67	β-Copaene		0.01			
68	*trans*-α-Bergamotene	0.21		0.14	0.20	0.25
69	*cis*-β-Farnesene			0.03		
70	α-Humulene[b]	0.03		0.03		0.03
71	*trans*-β-Farnesene	t	t		0.03	0.04 (
72	β-Santalene	t			t	
73	Germacrene D		t	0.01	0.01	0.02
74	Valencene	0.07	0.06	0.07	0.02	0.02
75	Bicyclogermacrene	0.02			0.02	0.04
76	*cis*-α-Bisabolene			0.02	0.02	0.02
77	*E,E*-α-Farnesene	0.02		0.02		
78	β-Bisabolene	0.40		0.21	0.31	0.39
79	δ-Cadinene		0.01	0.01		
80	*trans*-β-Santalol			0.01		
81	*epi*-Santalene	0.03			0.01	0.02
82	*cis*-Nerolidol acetate			0.02		
83	α-Bisabolol[b]	0.05		0.02	0.03	0.04
84	β-Sinensal		0.05	0.09		
85	α-Sinensal		0.02	0.02		
86	Nootkatone	0.02	0.02		0.01	
87	Tricosane	0.03				
	Classes[c]					
	Monoterpene hydrocarbons	**76.43[A]**	**94.16[C]**	**97.27[D]**	**88.66[B]**	**93.98[C]**
	Oxygenated monoterpenes	**20.37[E]**	**3.22[A]**	**1.78[A]**	**9.93[D]**	**4.58[C]**
	Sesquiterpenes	**1.14[E]**	**0.20[A]**	**0.67[B]**	**0.74[C]**	**0.98[D]**
	Others	**1.07[D]**	**2.31[E]**	**0.22[A]**	**0.56[C]**	**0.31[B]**

[a] Values (relative peak area percent) represent averages of 18 determinations for each cybrid and hybrid (nine for each collection year: 2006 and 2007), and 9 determination for parents (collection year 2007), (t = trace, $< 0.05\%$); [b] Co-elution with authentic sample. [C] Different letters in the same line represent significant differences at $p \leq 0.01$ by HSD Tukey test.

Conclusion

The development of plant molecular biology and, more recently, of plant genomics is a promising way for continuing genetic improvement of citrus. In the future, novel breeding technologies, such as genome resequencing, allele mining and genomic selection, and proteomics technology will integrate, and progressively replace, the traditional techniques for genotype characterization, germplasm screening and variety selection. The application of the genomic tools to plant breeding will enable breeders to plan novel varieties bearing pre-defined traits (breeding by design). Once the traits of interest have been identified in the genome, high-throughput molecular markers could be used to assemble the most favorable combination of characters in new varieties, then allowing the recovery and build-up of desirable crop phenotypes in novel hybrid species. However, the results of this study show, analogously to many similar ones, just how difficult it is to establish an inheritance mechanism related to the biosynthetic accumulation of secondary metabolites throughout various breeding methodologies of *Citrus* species. This is probably due to the complexity and genetic changeability of this genus [64, 70]. However, notwithstanding these intrinsic difficulties, it is desirable an ever-growing connection between the compositional studies of the secondary metabolism (metabolomic), comprising volatile and not volatile components, and the genomic studies, which could positively affect the future *Citrus* breeding programs.

Acknowledgment

Most of the work here described has been financially supported by Consiglio Nazionale delle Ricerche (C.N.R. - Rome) and partially in the context of the Project: "Risorse genetiche vegetali per la produzione di sostanze di interesse per la salute umana," Ministero Università e Ricerca Scientifica (MIUR – Rome). The authors also wish to thank the technical staff of both C.N.R. Institutes: ICB-CT (Mrs. Tonia Strano and Mr. Agatino Renda) and IBBR-PA (Mr. Vincenzo Marino) for their skillful technical assistance.

References

[1] Slafer, GA; Satorre, EH; Andrade, H. Increases in grain yield in bread wheat from breeding and associated physiological changes. In: Slafer, G, editor. Genetic improvement of field crops. New York, Marcel Dekker, 1994, pp. 1-67.

[2] Duvick, DN. The contribution of breeding to yield advances in maize (*Zea mays* L.) In: Donald LS, editor. Advances in Agronomy. Academic Press, 2005, pp 83-145.

[3] United States Department of Agriculture – National Agricultural Statistics Service (USDA-NASS), Field Crop Statistics, USDA-NASS, 2013.

[4] FAOStat (2011): http://faostat.fao.org/site/567/default.aspx#ancor.

[5] Aleza, P; Juarez, J; Ollitrault, P; Navarro, L. Polyembryony in non-apomictic citrus genotypes. *Ann. Bot.* 2010, *106*, 535-545.

[6] Almeida, WAB; Mourao Filho, F; Mendes, BMJ; Rodriguez, APM. *In vitro* organogenesis optimization and plantlet regeneration in *Citrus sinensis* and *C. limonia*. *Sci. Agr.* 2002, *59*, 35-40.

[7] Carimi, F. Somatic embryogenesis protocol: Citrus. Protocol for Somatic Embryogenesis. In: Jain, SM; Gupta, PK, editors. Woody Plants. Springer, The Netherlands, 2005, pp. 321-343.

[8] Duran-Vila, N; Ortega, V; Navarro, L. Morphogenesis and tissue cultures of three citrus species. *Plant Cell Tis. Org. Cult.* 1988, *16*, 125-133.

[9] Duran-Vila, N; Gogorcena, Y; Ortega, V; Ortiz, J; Navarro, L. Morphogenesis and tissue culture of sweet orange (*Citrus sinensis* L. Osb.): effect of temperature and photosynthetic radiation. *Plant Cell Tis. Org. Cult.* 1992, *29*, 11-18.

[10] Ghorbel, R; Navarro, L; Duran-Vila, N. Morphogenesis and regeneration of whole plants of grapefruit (*Citrus paradisi*), sour orange (*C. aurantium*) and alemow (*C. macrophilla*). *J. Hort. Sci. Biotechnol.* 1998, *73*, 323-327

[11] Goh, CJ; Sim, GE; Morales, CL; Loh, CS. Plantlet regeneration through different morphogenic pathways in pommelo tissue culture. *Plant Cell Tis. Org. Cult.* 1995, *43*, 301-303.

[12] Hidaka, T, Yamada, Y; Shicijo, T. *In vitro* differentiation of haploid plants by anther culture in *Poncirus trifoliate* (L.) Raf. *Japan J. Breed.* 1979, *29*, 248-258.

[13] Navarro, L. Citrus shoot tip grafting *in vitro*. In: Bajaj, YPS, editor. Biotechnology in Agriculture and Forestry - Vol. 18. Berlin, Heidelberg, Springer, 1992, pp. 327-338.

[14] Navarro, L, Olivares-Fuster, O; Juarez, J; Aleza, P; Pena, JA; Ballester-Olmos, JF; Cervera, M; Fagoaga, C; Duran-Vila, N; Pena, L. Application of biotechnology to Citrus improvement in Spain. *Acta Hort.* 2004, *632*, 221-234.

[15] Nito, N; Iwamasa, M. *In vitro* plantlet formation from juice vesicle callus of Satsuma (*Citrus unshiu* Marc.). *Plant Cell Tis. Org. Cult.* 1990, *20*, 137-140.

[16] Obukosia, SD; Waithaka, K. Nucellar embryo culture of *Citrus sinensis* L. and *Citrus limon* L. *African Crop Sci. J.* 2000, *8*, 109-116.

[17] Perez-Molphe-Bach, E; Ochoa-Alejo, N. *In vitro* plant regeneration of Mexican lime and mandarin by direct organogenesis. *HortSci.* 1977, *32*, 931-934.

[18] Sim, GE; Goh, CJ; Loh, CS. Micropropagation of *Citrus mitis* Blanco. Multiple bud formation from shoot and root explants in the presence of 6-benzylaminopurine. *Plant Sci.* 1989, *59*, 203-210.

[19] Grosser, JW; Gmitter, FG. Protoplast fusion and citrus improvement. *Plant Breeding Rev.* 1990, *8*, 339-374.

[20] Davey, MR; Anthony, P; Power, JB; Lowe, KC. Plant protoplasts: status and biotechnological perspectives. Research review paper. *Biotech. Adv.* 2005, *25*, 131-171.

[21] Khan, IA. Citrus genetics, breeding and biotechnology. Wallingford, UK, CABI Publishing, 2007.

[22] Grosser, JW; Gmitter, FG. Somatic hybridization of Citrus with wild relatives for germplasm enhancement and cultivar development. *HortScience,* 1990, *25(2)*, 147-151.

[23] Guo, WW; Prasad, D; Cheng, YJ; Serrano, P; Deng, XX; Grosser, JW. Targeted cybridization in *Citrus:* transfer of Satsuma cytoplasm to seedy cultivars for potential seedlessness. *Plant Cell Rep.* 2004, *22*, 752-758.

[24] Grosser, JW; Ollitrault, P; Olivares-Fuster, O. Somatic hybridization in citrus: an effective tool to facilitate variety improvement. *In Vitro Cell. Dev. Biol. Plant.* 2000, *36*, 434-449.

[25] Grosser, JW; Gmitter, FJ. Protoplast fusion for production of tetraploid and triploid: applications for scion and rootstock breeding in citrus. *Plant Cell Tiss. Organ Cult.* 2011, *104*, 343-357.

[26] Ollitrault, P; Guo, W; Grosser JW. Recent advances and evolving strategies in *Citrus* somatic hybridization. In: Khan, IH, editor. *Citrus genetics, breed and biotechnology.* Wallingford, UK, CABI Publishing, 2007, pp. 235-260.

[27] Syvertsen, JP; Lee, LS; Grosser, JW. Limitations on growth and net gas exchange of diploid and tetraploid *Citrus* rootstock cultivars grown at elevated CO_2. *J. Am. Soc. Hort. Sci.* 2000, *125*, 228-234.

[28] Morillon, R; Allario, T; Brumos, J; Colmenero, JM; Iglesias, D; Pina, JA,; Tadeo, F; Talon, M; Navarro, L; Ollitrault, P. Tetraploid *Citrus limonia* rootstocks are more tolerant to water deficit than parental diploids and present large gene expression changes in roots. In: Proceedings of 28th International Horticulture Congress, Lisbon, Portugal, 2010.

[29] Ohgawara, T; Kobayashi, S; Ohgawara, E; Uchimiya, H; Ishii, S. Somatic hybrid plants obtained by protoplast fusion between *Citrus sinensis* and *Poncirus trifoliata. Theor. Appl. Genet.* 1985, *71*, 1-4.

[30] Abbate, L; Tusa, N; Fatta Del Bosco, S; Strano, T; Renda, A; Ruberto, G. Genetic improvement of *Citrus* fruits: new somatic hybrids from *Citrus sinensis* (L) Osb and *Citrus limon* (L) Burm. F. *Food Res. Int.* 2012, *48*, 284-290.

[31] Gancel, AL; Olle, D; Ollitrault, P; Luro, F; Brillouet, JM. Leaf and peel volatile compounds of an interspecific *Citrus* somatic hybrid (*Citrus aurantium* Christm + *Citrus paradisi* Macfayden). *Flavour Fragr. J.* 2002, *17*, 416-424.

[32] Gancel, AL; Ollitrault, P; Froelicher, Y; Tomi, F; Jacquemond, C; Luro, F; Brillouet, JM. Leaf volatile compounds of seven citrus somatic tetraploid hybrids sharing willow leaf mandarin (*Citrus deliciosa* Ten.) as their common parent. *J. Agric. Food Chem.* 2003, *51*, 6006-6013.

[33] Gancel, AL; Ollitrault, P; Froelicher, Y; Tomi, F; Jacquemond, C; Luro, F; Brillouet, JM. Citrus somatic allotetraploid hybrids exhibit a differential reduction of leaf sesquiterpenoid biosynthesis compared with their parents. *Flavour Fragr. J.* 2005, *20*, 626-632.

[34] Gancel, AL; Ollitrault, P; Froelicher, Y; Tomi, F; Jacquemond, C; Luro, F; Brillouet, JM. Leaf volatile compounds of six Citrus somatic allotetraploid hybrids originating from various combination of lime, lemon, citron, sweet orange, and grapefruit. *J. Agric. Food Chem.* 2005, *53*, 2224-2230.

[35] Bassene, JB; Berti, L; Costantino, G; Carcouet, E; Kamiri, M; Tomi, F; Dambier, D; Ollitrault, P; Froelicher, Y. Inheritance of characters involved in fruit quality in a citrus interspecific allotetraploid somatic hybrid. *J. Agric. Food Chem.* 2009, *57*, 5065-5070.

[36] Bassene, JB; Froelicher, Y; Dhuique-Mayer, C; Mouhaya, W; Ferrer, RM; Ancillo, G; Morillon, R; Navarro, L; Ollitrault, P. Non-additive phenotypic and transcriptomic inheritance in a citrus allotetraploid somatic hybrid between *C. reticulata* and *C. limon*: the case of pulp carotenoid biosynthesis pathway. *Plant Cell Rep.* 2009, *28*, 1689-1697.

[37] Louzada, ES; Grosser, JW; Gmitter, FJ. Intergeneric somatic hybridization of sexually incompatible *Citrus sinensis* and *Atalantia ceylanica*. *Plant Cell Rep*. 1993, *12*, 687-690.

[38] Sherraf, I; Tizroutine, S; Chaput, MH; Allot, M; Mussio, I; Sihachakr, D, Rossignol, L; Ducreux, G. Production and characterization of intergeneric somatic hybrids through protoplast electrofusion between potato (*Solanum tuberosum*) and *Lycopersicon pennellii*. *Plant Cell Tiss. Org. Cul*t 1994, *37*, 137-144.

[39] Fatta Del Bosco, S; Tusa, N; Conicella, C. Microsporogenesis in a *Citrus* interspecific somatic hybrid and its fusion parents. *Heredity* 1999, *83*, 373-377.

[40] Dambier, D; Benyahia, H; Pensabene Bellavia, G; Kaçar, YA; Froelicher, Y; Belfalah, Z; Lhou, B; Handaji, N; Printz, B; Morillon, R; Yesiloglu, T; Navarro, L; Ollitrault, P. Somatic hybridization for citrus rootstock breeding: an effective tool to solve some important issues of the Mediterranean citrus industry. *Plant Cell Rep*. 2011, *30*, 883-900.

[41] Esen, A; Soost, RK. Unexpected triploids in citrus: their origin, identification and possible use. *J. Hered*. 1971, *62*, 329-333.

[42] Cameron, JW; Burnett, RH. Use of sexual tetraploid seed parents for production of triploid citrus hybrids. *HortSci* 1978, *13*, 167-169.

[43] Tusa, N; Fatta Del Bosco, S; Nardi, L; Lucretti, S. Obtaining triploid plants by crossing *Citrus limon cv "Femminello"* 2n x 4n allotetraploid somatic hybrids, *Proc. Int. Soc. of Citricult.* 1996, 133-136.

[44] Fatta Del Bosco, S; Abbate, L; Tusa, N; Strano, T; Renda, A; Ruberto, G. Genetic improvement of *Citrus* fruit: the essential oil profiles in a *Citrus limon* backcross progeny derived from somatic hybridization. *Food Res. Int.* 2013, *50*, 344-350.

[45] Dudits, D; Fejer, G; Hadlaezdy, C; Koncz, C; Lazar, GB; Horvfith, G. Intergeneric gene transfer mediated by plant protoplast fusion. *Mol. Gen. Genet*. 1980, *179*, 283-288.

[46] Li, YG; Stoutjestiik, PA; Delves, AC; Larkin, PJ. Somatic hybridization for plants improvement. Dordrecht, Kluwer Academic Publishers, 1999.

[47] Liu, JH; Xu, XY; Deng, XX. Intergeneric somatic hybridization and its application to crop genetic improvement. *Plant Cell Tiss. Organ Cult*. 2005, *82*, 19-44.

[48] Ohgawara, T; Kobayashi, S; Ishii, S; Yoshinaga, K; Oiyama, I. Somatic hybridization in *Citrus*: navel orange (*C. sinensis* Osb.) and grapefruit (*C. paradisi* Macf.). *Theor Appl. Genet*. 1989, *78*, 609-612.

[49] Tusa, N; Grosser, JW; Gmitter, FG. Plant regeneration of "Valencia" sweet orange, "Femminello" lemon and the interspecific somatic hybrid following protoplast fusion. *J. Amer. Soc. Hort. Sci*. 1990, *115(6),* 1043-1046.

[50] Grosser, JW; Gmitter, FG; Sesto, F; Deng, XX; Chandler, JL. Six new somatic *Citrus* hybrids and their potential for cultivar improvement. *J. Amer. Soc. Hort. Sci*. 1992, *117*, 169-173.

[51] Louzada, ES; Grosser, JW, Gmitter, FG; Nielsen, B; Deng, XX; Tusa, N; Chandler, JL. Eight new somatic hybrid *Citrus* rootstocks with potential for improved disease tolerance. *HortScience* 1992, *27*, 1033-1036.

[52] Saito, W; Ohgawara, T; Shimizu, J; Ishii, S; Kobayashi, S. *Citrus* cybrid regeneration following cell fusion between nucellar cells and mesophyll cells. *Plant Sci.* 1993, *88*, 195-201.

[53] Grosser, JW; Gmitter, FG; Tusa, N; Reforgiato Recupero, G; Cucinotta, P. Further evidence of a cybridization requirement for plant regeneration from citrus leaf protoplasts following somatic fusion. *Plant Cell Rep*. 1996, 15, 672-676.

[54] Moriguchi, T; Hidaka, T; Omura, M; Motomura, T; Akihama, T. Genotypes and parental combination influence efficiency of cybrid induction in *Citrus* by electro fusion. *HortScience* 1996, *31*, 275-278.

[55] Moreira, CD; Chase, CD; Gmitter, FG; Grosser, JW. Inheritance of organelle genomes in citrus somatic cybrids. *Mol Breed*. 2000, *6*, 401-405.

[56] Cabasson, CM; Luro, E; Ollitrault, P; Grosser, JW. Non-random inheritance of mitochondrial genomes in *Citrus* hybrids produced by protoplast fusion. *Plant Cell Rep*. 2001, *20*, 604-609.

[57] Guo, WW; Xiao, SX; Deng, XX. Somatic cybrid production via protoplast fusion for citrus improvement. *Sci. Hortic* 2013, *163*, 20-26.

[58] Xiao, SX; Biswas, MK; Li, MY; Deng, XX; Xu, Q; Guo WW. Production and molecular characterization of diploid and tetraploid somatic cybrid plants between male sterile Satsuma mandarin and seedy sweet orange cultivars. *Plant Cell Tiss. Org. Cult*. 2014, *116*, 81-88.

[59] Koussevitzky, S; Nott, A; Mockler, TC; Hong, FX; Sachetto-Martins, G; Surpin, M; Lim, J; Mitter, R; Chory, J. Signals from chloroplasts converge to regulate nuclear gene expression. *Science* 2007, *316*, 715-719.

[60] Yang, JJ; Zhang, M; Yu, J. Mitochondrial retrograde regulation tuning fork in nuclear genes expression of higher plants. *J. Genet. Genomics* 2008, *35*, 65-71.

[61] Wang, L; Pan, ZY; Guo, WW. Proteomic analysis of leaves from a diploid cybrid produced by protoplast fusion between Satsuma mandarin and pummelo. *Plant Cell Tiss. Org. Cult*. 2010, *103*, 165–174.

[62] Fanciullino, AL; Gancel, AL; Froelicher, Y; Luro, F; Ollitrault, P; Brillouet, JM. Effects of nucleo-cytoplasmic interactions on leaf volatile compounds from citrus somatic diploid hybrids. *J. Agric. Food Chem.* 2005, *53*, 4517-4523.

[63] Bassene, JB; Berti, L; Carcouet, E; Dhuique-Mayer, C; Fanciullino, AL; Bouffin, J; Ollitrault, P; Froelicher, Y. Influence of mitochondria origin on fruit quality in a Citrus cybrid. *J. Agric. Food Chem*. 2008, *56*, 8635-9640.

[64] Guo, WW; Cheng, YJ; Chen, CL; Deng, XX. Molecular analysis revealed autotetraploid, diploid and tetraploid cybrid plants regenerated from an interspecific somatic fusion in *Citrus*. *Sci Hort* 2006, *108*, 162-166.

[65] Guo, WW; Wu, RC; Cheng, YJ; Deng, XX. Production and molecular characterization of *Citrus* intergeneric somatic hybrids between red tangerine and citrange. *Plant Breed*. 2007, *126*, 72-76.

[66] Fang, DQ; Roose, ML. Identification of closely related citrus cultivars with inter-simple sequence repeat markers. *Theor. Appl. Genet.* 1997, *95*, 408-417.

[67] Coletta Filho, HD; Machado, MA; Targon, MLPN; Moreira, MCPQDG, Pompeu, J. Analysis of the genetic diversity among mandarins (*Citrus* spp.) using RAPD markers. *Euphytica* 1998, *102*, 139-144.

[68] Federici, CT; Fang, DQ; Scora, RW; Roose, ML. Phylogenetic relationships within the genus Citrus (Rutaceae) and related genera as revealed by RFLP and RAPD analysis. *Theor. Appl. Genet.* 1998, *96(6/7),* 812-822.

[69] Luro, F; Rist, D; Ollitrault, P. Evaluation of genetic relationships in Citrus genus by means of sequence tagged microsatellites. *Acta Horticult.* 2001, *546*, 237-242.

[70] Barkley, NA; Roose, ML; Krueger, RR; Federici, CT. Assessing genetic diversity and population structure in a citrus germplasm collection utilizing simple sequence repeat markers (SSRs). *Theor. Appl. Genet.* 2006, *112(8),* 1519-1531.

[71] Garcia-Lor, A. Comparative contribution of InDels and SSRs markers to decipher the interspecific structure of the cultivated Citrus genetic diversity; perspective for association genetic studies. *Mol. Gen. Genet.* 2011, *287(1),* 77-94.

[72] Ollitrault, P; Terol, J; Garcia-Lor, A, Bérard, A; Chauveau, A, Froelicher, Y; Belzile, C; Morillon, R; Navarro, L; Brunel, D; Talon, M. SNP mining in C. clementina BAC end sequences; transferability in the *Citrus* genus (Rutaceae), phylogenetic inferences and perspectives for genetic mapping. *BMC Genomics* 2012, *13*, 13.

[73] Kijas, JMH; Thomas, MR; Fowler, JCS; Roose, ML. Integration of trinucleotide microsatellites into a linkage map of Citrus. *Theor. Appl. Genet.* 1997, *94*, 701–706.

[74] Ling, P; Duncan, LW; Deng, Z; Dunn, D; Hu, X; Huang, S, Gmitter, FG Jr. Inheritance of citrus nematode resistance and its linkage with molecular markers. *Theor. Appl. Genet.* 2000, *100,* 1010–1101.

[75] Sankar, AA; Moore, GA. Evaluation of inter-simple sequence repeat analysis for mapping in Citrus and extension of the genetic linkage map. *Theor. Appl. Genet.* 2001, *102*, 206–214.

[76] Gulsen, O; Uzun, A; Canan, I; Seday, U; Canihos, E. A new citrus linkage map based on SRAP, SSR, ISSR, POGP, RGA and RAPD markers. *Euphytica* 2010, *173*, 265–277.

[77] Nicolosi, E; Deng, ZN; Gentile, A; La Malfa, S; Continella, G; Tribulato, E. Citrus phylogeny and genetic origin of important species as investigated by molecular markers. *Theor. Appl. Genet.* 2000, *100*, 1155–1166.

[78] Scarano,, MT; Tusa, N; Abbate, L; Lucretti, S; Nardi, L; Ferrante, S. Flow cytometry, SSR and modified AFLP markers for the identification of zygotic plantlets in backcrosses between 'Femminello' lemon cybrids (2n and 4n) and a diploid clone of 'Femminello' lemon (*Citrus limon* L. Burm. F.) tolerant to mal secco disease. *Plant Sci.* 2003, *164*, 1009-1017.

[79] Cheng, Y; De Vicente, MC; Meng, H; Guo, W; Tao, N; Deng, X. A set of primers for analyzing chloroplast DNA diversity in Citrus and related genera. *Tree Physiol.* 2005, *25*, 661–672.

[80] Uzun, A; Yesiloglu, T; Aka-Kacar, Y; Tuzcu, O; Gulsen, O. Genetic diversity and population structure within Citrus and related genera based on sequence related amplified polymorphism markers (SRAPs). *Sci Hortic* 2009, *121*, 306–312.

[81] Palmieri, DA; Novelli, VM; Bastianel, M; Cristofani-Yaly, M; Astúa-Monge, G; Carlos, EF; Carlos de Oliveira, A; Machado, MA. Frequency and distribution of microsatellites from ESTs of citrus. *Genet. Mol. Biol.* 2007, *30*, 1009–1018.

[82] Holton, TA; Christopher, JT; McClure, L; Harker, N; Henry, RJ. Identification and mapping of polymorphic SSR markers from expressed gene sequences of barley and wheat. *Mol. Breed.* 2002, *9*, 63–71.

[83] Zane, L; Bargelloni, L; Patarnello, T. Strategies for microsatellite isolation. *Mol. Ecol.* 2002, *11*, 1–16.

[84] Jannati, M; Fotouhi, R; Pourjan, A; Salehi, AZ. Genetic diversity analysis of Iranian citrus varieties using microsatellite (SSR) based markers. *J. Hortic. For.* 2009, *1*, 120–125.

[85] Fu, J; Peng ZJ; Cai XD; Guo WW. Regeneration and molecular characterization of interspecific somatic hybrids between Satsuma mandarin and two seedy sweet oranges for scion improvement. *Plant Breed.* 2010, 01773, doi:10.1111/j.1439-0523.

[86] Amar, MH; Biswas, MK; Zhang, Z; Guo, WW. Exploitation of SSR, SRAP and CAPS-SNP markers for genetic diversity of Citrus germplasm collection. *Scientia Horticult.* 2011, *128*, 220–227.

[87] Tautz, D. Hypervariability of simple sequences as a general source for polymorphic markers. *Nucleic Acid Res.* 1989, *17*, 6463-6471.

[88] Estoup, A; Jarne, P; Cornuet, JM. Homoplasy and mutation model at microsatellite loci and their consequences for population genetics analysis. *Mol Ecol.* 2002, *11*, 1591–1604.

[89] Paetkau D; Waits, LP; Clarkson, PL; Craighead, L; Strobeck, C. An empirical evaluation of genetic distance statistics using microsatellite data from bear (Ursidae) populations. *Genetics* 1997, *147*, 1943–1957.

[90] Weising, K; Gardner, RC. A set of conserved PCR primers for the analysis of simple sequence repeat polymorphisms in chloroplast genomes of dicotyledonous angiosperms. *Genome* 1999, *42*, 9–19.

[91] Xu, DH; J. Abe, JY; Gai, N; Shimamoto, Y. Diversity of chloroplast DNA SSRs in wild and cultivated soybeans: evidence for multiple origins of cultivated soybean. *Theor. Appl. Genet.* 2002, *105*, 645–653.

[92] Vogel, M; Bänfer, G; Moog, U; Weising, K. Development and characterization of chloroplast microsatellite markers in Macaranga (Euphorbiaceae). *Genome* 2003, *46*, 845–857.

[93] Provan, J; Soranzo, N.; Wilson, NJ; Goldstein, DB; Powell, W. A low mutation rate for chloroplast microsatellites. *Genetics* 1999, *153*, 943–947.

[94] Ishii, T; McCouch; SR. Microsatellites and microsynteny in the chloroplast genomes of *Oryza* and eight other Gramineae species. *Theor. Appl. Genet.* 2000, *100*, 1257–1266.

[95] Besnard, G; Khadari, B; Baradat, P; Bervillé, A. *Olea europaea* (Oleaceae) phylogeography based on chloroplast DNA polymorphism. *Theor. Appl. Genet.* 2002, *104*, 1353–1361.

[96] Palme, AE; Vendramin, GG. Chloroplast DNA variation, postglacial recolonization and hybridization in hazel, *Corylus avellana*. *Mol. Ecol.* 2002, *9*, 1769–1779.

[97] Vettori, C; Vendramin, GG; Anzidei, M; Pastorelli, R; Paffetti, D; Giannini, R. Geographic distribution of chloroplast variation in Italian populations of beech (*Fagus sylvatica* L.). *Theor. Appl. Genet.* 2004, *109*, 1–9.

[98] Cheng, YJ; Guo, WW; Deng, XX. Molecular characterization of cytoplasmic and nuclear genomes in phenotypically abnormal Valencia orange (*Citrus sinensis*) + Meiwa kumquat (*Fortunella crassifolia*) intergeneric somatic hybrids. *Plant Cell Rep.* 2003, *21*, 445–451.

[99] Deng, Z; La Malfa, S; Xie, Y; Xiong, X; Gentile, A. Identification and evaluation of chloroplast uni- and trinucleotide sequence repeats in citrus. *Scientia Horticult.* 2007, *111*, 186–192.

[100] Green, RM; Vardi, A; Galun, E. The plastome of Citrus. Physical map, variation among Citrus cultivars and species and comparison with related genera. *Theor. Appl. Genet.* 1986, *72*, 170–177.

[101] Yamamoto, M; Kobayashi, S; Nakamura, Y; Yamada, Y. Phylogenic relationships of citrus revealed by RFLP analysis of mitochondrial and chloroplast DNA. *Jap. J. Breed.* 1993, *43*, 355-365.

[102] Moreira, CD; Gmitter, FG Jr; Grosser, JW; Huang, S; Ortega, VM; Chase, CD. Inheritance of organelle DNA sequences in a Citrus-Poncirus intergeneric cross. *J. Hered.* 2002, *93*, 174–178.

[103] Lotfy, S, Luro, F; Carreel, F; Froelicher, Y; Rist, D; Ollitrault, P. Application of cleaved amplified polymorphic sequence method for analysis of cytoplasmic genome among Aurantioideae intergeneric somatic hybrids. *J. Am. Soc. Hortic. Sci.* 2003, *128*, 225–230.

[104] Gray, MW; Burger, G; Lang, BF. Mitochondrial evolution. *Science* 1999, *283*, 1476–1481.

[105] Muse, SV. Examining rates and patterns of nucleotide substitution in plants. *Plant Mol. Biol.* 2000, *42*, 25–43.

[106] Froelicher, Y; Mouhaya, W; Bassene, JB; Costantino, G; Kamiri, M; Luro, F; Morillon, R, Ollitrault, P. New universal mitochondrial PCR markers reveal new information on maternal citrus phylogeny. *Tree Genet. Gen.* 2011, *7*, 49-61.

[107] Ruan, C. Germplasm-regression-combined marker-trait association identification in plants. *Afr. J. Biotechnol.* 2010, *9*, 573–580.

[108] Morrell, PL; Buckler, ES; Ross-Ibarra, J. Crop genomics: advances and applications. *Nat. Rev. Genet.* 2012, *13*, 85-96.

[109] Kumar, S; Garrick, DJ; Bink, MC; Whitworth, C; Chagné, D; Volz, RK. Novel genomic approaches unravel genetic architecture of complex traits in apple. *BMC Genomics.* 2013, *14*, 393.

[110] Ganal, MW; Altmann, T; Roder, MS. SNP identification in crop plants. *Curr. Opin. Plant Biol.* 2009, *12*, 211-217.

[111] Schmid, KJ; Sorensen, TR; Stracke, R; Torjek, O; Altmann, T; Mitchell-Olds, T, Weisshaar, B. Large-scale identification and analysis of genome-wide singlem nucleotide polymorphisms for mapping in *Arabidopsis thaliana*. *Genome Res.* 2003, *13*, 1250-1257.

[112] Yu, H; Xie, W; Li, J; Zhou, F; Zhang, QA. Whole-genome SNP array (RICE6K) for genomic breeding in rice. *Plant Biotechnol J.* 2014, *12*, 28-37.

[113] Hyten, DL; Song, QJ; Choi, IY; Yoon, MS; Specht, JE; Matukumalli, LK; Nelson, RL; Shoemaker, RC; Young, ND; Cregan, PB. High-throughput genotyping with the Golden Gate assay in the complex genome of soybean. *Theor. App. Genetics.* 2008, *116*, 945-952.

[114] Sim, SC; van Deynze, A; Stoffel, K; Douches, DS; Zarka, D; Ganal, MW; Chetelat, RT, Hutton, SF; Scott, JW; Gardner, RG; Panthee, DR; Mutschler, M; Myers, JR; Francis, DM. High-density SNP genotyping of tomato (*Solanum lycopersicum* L.) reveals patterns of genetic variation due to breeding. *PLoS ONE.* 2012, *7*, e45520.

[115] Cavanagh, CR; Chao, S; Wang, S; Huang, BE; Stephen, S; Kiani, S; Forrest, K; Saintenac, C; Brown-Guedira, GL; Akhunova, A; See, D; Bai, G; Pumphrey, M; Tomar, L; Wong, D; Kong, S; Reynolds, MP; Lopez de Silva, M; Bockelman, H;

Talbert, L; Anderson, JA; Dreisigacker, S; Baenziger, S; Carter, A; Korzun, V; Morrell, PL; Dubcovsky, J; Morrell, MK; Sorrels, ME; Hayden, MJ; Akhunov ED. Genome-wide comparative diversity uncovers multiple targets of selection for improvement in hexaploid wheat landraces and cultivars. *Proc. Natl. Acad. Sci. USA* 2013, *110*, 8057-8062.

[116] Chagné, D; Crowhurst, RN; Troggio, M; Davey, MW; Gilmore, B; Lawley, C; Vanderzande, S, Hellens, RP; Kumar, S; Cestaro, A; Velasco, R; Main, D; Rees, JD; Iezzoni, A; Mockler, T; Wilhelm, L; Van de Weg, E; Gardiner, SE; Bassil, N; Peace, C. Genome-wide SNP detection, validation, and development of an 8K SNP array for apple. *PLoS ONE* 2012, *7*, e31745.

[117] Ahmad, R; Parfitt, DE; Fass, J; Ogundiwin, E; Dhingra, A; Gradziel, TM; Lin, D; Joshi, NA; Martinez-Garcia, PJ; Crisosto, CH. Whole genome sequencing of peach (*Prunus persica* L) for SNP identification and selection. *BMC Genomics* 2011, *12*, 569.

[118] Fujii, H; Shimada, T; Nonaka, K; Kita, M; Kuniga, T; Endo, T; Ikoma, Y; Omura, M. High-throughput genotyping in citrus accessions using an SNP genotyping array. *Tree Genet. Gen.* 2013, *9*, 145–153.

[119] Murakami, A; Nakamura, Y; Torikai, K; Tanaka, T; Koshiba, T; Koshimizu, K; Kuwahara, S; Takahashi, Y; Ogawa, K; Yano, M; Tokuda, H; Nishini, H; Mimaki, Y; Sashida, Y; Kitanaka, S; Ohigashi, H. Inhibitory effect of citrus nobiletin on phorbol ester-induced skin inflammation, oxidative stress, and tumor-promotion in mice. *Cancer Res.* 2000, *60*, 5059-5066.

[120] Tsushima, M; Maoka, T; Katsuyama, M; Kozuka, M; Matsuno, T; Tokuda, H; Nishino, H; Iwashima, A. Inhibitory effect of natural carotenoids on Epstein–Barr virus activation activity of a tumor promoter in Raji cells. A screening study for anti-tumor promoters. *Biol. Pharm. Bull.* 1995, *18*, 227–233.

[121] Lam, LKT; Hasegawa, S. Inhibition of benzo-[α]-pyrene-induced fore-stomach neoplasia in mice by citrus limonoids. *Nutr. Canc.* 1989, *12*, 43-47.

[122] Lam, LKT; Li, Y; Hasegawa, S. Effects of citrus limonoids on glutathione S-transferase activity in mice. *J. Agric. Food Chem.* 1989, *37*, 878–880.

[123] Sugiyama, A; Omura, M; Matsumoto, H; Shimada, T; Fujii, H; Endo, T; Shimizu, T, Nesumi, H; Ikoma, Y. Quantitative trait loci (QTL) analysis of carotenoid content in Citrus fruit. *J. Japan Soc. Hort. Sci.* 2011, *80*, 136–144.

[124] Edwards, D; Batley, J. Plant genome sequencing: applications for crop improvement. *Plant Biotechn. J.* 2010, *8(1),* 2-9.

[125] Chen, D; Ahlford, A; Schnorrer, F; Kalchhauser, I; Fellner, M; Viragh, E; Kiss, I; Syvanen, AC; Dickson, BJ. High-resolution, high-throughput SNP mapping in *Drosophila melanogaster*. *Nat. Methods* 2008, *5(4),* 323-329.

[126] McCouch, SR; Zhao, K; Wright, M; Tung, C; Ebana, K; Thomson, M; Reynolds, A; Wang, D; DeClerck, G; Ali, ML; McClung, A; Eizenga, G; Bustamante, C. Development of genome-wide SNP assays for rice. *Breed Sci.* 2010, *60(5),* 524-535.

[127] Tung, CW; Zhao, K; Wright, MH; Liakat Ali, M; Jung, J; Kimball, J; Tyagi, W; Thomson, MJ; McNally, K; Leung, H; Kim, H; Ahn, SN; Reynolds, A; Scheffer, B; Eizengam G; McClung, A; Bustamante, C; McCouch, SR. Development of a research platform for dissecting phenotype–genotype associations in rice (*Oryza* spp.). *Rice* 2010, *3*, 205–217.

[128] Rostoks, N; Ramsay, L; Mackenzie, K; Cardle, L; Bhat, PR; Roose, ML; Svensson, JT; Stein, N; Varshney, RK; Marshall, DF; Graner, A; Close, TJ; Waugh, R. Recent history of artificial outcrossing facilitates whole-genome association mapping in elite inbred crop varieties. *Proc. Natl. Acad. Sci. USA* 2006, *103*, 18656–18661.

[129] Hyten, DL; Song, Q; Choi, IY; Yoon, MS; Specht, JE; Matukumalli, LK; Nelson, RL; Shoemaker, RC; Young, ND; Cregan, PB. High-throughput genotyping with the GoldenGate® assay in the complex genome of soybean. *Theor. Appl. Genet.* 2008, *116*, 945–952.

[130] Pavy, N; Pelgas, B; Beauseigle, S; Blais, S; Gagnon, F; Gosselin, I; Lamothe, M; Isabel, N; Bousquet, J. Enhancing genetic mapping of complex genomes through the design of highly multiplexed SNP arrays: application to the large and unsequenced genomes of white spruce and black spruce. *BMC Genom* 2008, *9*, 21.

[131] Eckert, AJ; Pande, B, Ersoz, ES; Wright, MH; Rashbrook, VK; Nicolet, CM; Neale, DB. High-throughput genotyping and mapping of single nucleotide polymorphisms in loblolly pine (*Pinus taeda* L.). *Tree Genet. Genom.* 2009, *5,* 225–234.

[132] Uchiyama, K; Ujino-Ihara, T; Ueno, S; Taguchi, Y; Futamura, N; Shinohara, K; Tsumura, Y. Single nucleotide polymorphism in *Cryptomeria japonica*: their discovery and validation for genome mapping and diversity studies. *Tree Genet Genom*. 2012, *8*, 1213-1222.

[133] De Lorenzis, G; Chipashvili, R; Failla, O; Maghradze, D. Study of genetic variability in *Vitis vinifera* L. germplasm by high-throughput Vitis18kSNP array: the case of Georgian genetic resources. *BMC Plant Biol.* 2015, *15*, 154.

[134] Ruberto, G; Renda, A; Tringali, C; Napoli, EM; Simmonds, MSJ. Citrus limonoids and their semisyntheitic derivatives as antifeedant agents against *Spodoptera frugiperda* larvae. A structure-activity relationship study. *J. Agric. Food Chem.* 2002, *50*, 6766-6774.

[135] Kato, M. Mechanism of carotenoid accumulation in Citrus fruit. *J. Japan Soc. Hort. Sci.* 2012, *81(3)*, 219-233.

[136] Ikoma, Y; Matsumoto, H; Kato, M. The characteristics of carotenoids biosynthesis in Citrus fruit. *Japan Agric. Res. Quart.* 2014, *48(1),* 9-16.

[137] Ruberto, G. Analysis of volatile components of Citrus fruit essential oils. In: Jackson, JF; Linskens, HF; Inman, RB. Editors. Molecular Methods of Plant Analysis – Analysis of Taste and Aroma. Heidelberg, Germany, Springer-Verlag, 2002, pp. 123-157.

[138] Tranchida, PQ; Bonaccorsi, I; Dugo, P; Mondello, L; Dugo, G. Analysis of Citrus essential oils: State of the art and future perspectives. A review. *Flavour Fragr. J.* 2012, *27*, 98–123.

[139] Kaur, J; Kaur, G. An insight into the role of citrus bioactive in modulation of colon cancer. *J. Funct. Foods* 2015, *13*, 239-261.

[140] Khan, MK; E-Huma, Z; Dangles, O. A comprehensive review on flavanones, the major citrus polyphenols. *J. Food Comp. Anal.* 2014, *33*, 85-104.

[141] Gonzalez-Molina, E; Dominguez-Perles, R; Moreno, DA; Garcia-Viguera, C. Natural bioactive compounds of *Citrus limon* for food and health. *J. Pharm. Biomed. Anal.* 2010, 51, 327-345.

[142] Arroo, RRJ; Beresford, K; Bhambra, AS; Boarder, M; Budriesi, R; Cheng, Z; Micucci, M; Rupareila, KC; Surichan, S; Androutsopoulos, VP. Phytoestrogens as natural

prodrugs in cancer prevention: towards a mechanistic model. *Phytochem. Rev.* 2014, *13*, 853-866.

[143] Ignat, I; Volf, I; Popa, VI. A critical review of methods for characterization of polyphenolic compounds in fruits and vegetables. *Food Chem.* 2011, *126*, 1821-1835.

[144] Siracusa, L; Ruberto, G. Plant polyphenol profiles as a tool for traceability and valuable support to biodiversity. In: Watson, RR, editor. Polyphenol in Plants – Isolation, Purification and Extract Preparation, Elsevier Inc., 2014, pp. 15-33.

[145] Luykx, DMAM; van Ruth, SM. An overview of analytical methods for determining the geographical origin of food products. *Food Chem.* 2008, *107*, 897-911.

[146] Macel, M; van Dam, NM; Keurentjes, JJB. Metabolomics: the chemistry between ecology and genetics. *Mol. Ecol. Res.* 2010, *10*, 583-593.

[147] Yonekura-Sakakibara, K; Saito, K. Functional genomics for plant natural product biosynthesis. *Nat. Prod. Rep.* 2009, *26*, 1466-1487.

[148] Zhao, H; Guo, B; Wei, Y; Zhang, B. Effects of wheat origin, genotype, and their interaction on multielement fingerprints for geographical traceability. *J. Agric. Food Chem.* 2012, *60*, 10957-10962.

[149] Ruberto, G; Biondi, D; Piattelli, M; Rapisarda, P; Starrantino, A. Profiles of essential oils of new *Citrus* hybrids. *Flavour Fragr. J.* 1993, *8*, 179-184.

[150] Ruberto, G; Biondi, D; Piattelli, M; Rapisarda, P; Starrantino, A. Essential oil of the new *Citrus* hybrid, *Citrus clementina x C. limon. J. Essent. Oil Res.* 1994, *6*, 1-8.

[151] Ruberto, G; Renda, A; Piattelli, M; Rapisarda, P; Starrantino, A. Essential oil of two new pigmented Citrus hybrids, *Citrus clementina x C. sinensis. J. Agric. Food Chem.* 1997, *45*, 467-471.

[152] Ruberto, G; Biondi, D; Rapisarda, P; Renda, A; Starrantino, A. Essential oil of "Cami" a new Citrus hybrids. *J. Agric. Food Chem.* 1997, *45*, 3206-3210.

[153] Ruberto, G; Starrantino, A; Rapisarda, P. Citrus Improvement – Chemical and Genetic Aspects. In Pandalai, SG, editor. Recent Research Developments in Agricultural & Food Chemistry, Research Signpost Vol. 3, 1999, pp. 445-470.

[154] Ruberto, G; Rapisarda, P. Essential oils of new pigmented Citrus hybrids: *Citrus sinensis* L. Osbeck x *C. clementina* Hort ex Tanaka. *J. Food Sci.* 2002, *67*, 2778-2780.

[155] Tusa, N; Abbate, L; Renda, A; Ruberto, G. Polyphenols distribution in juices from Citrus allotetraploid somatic hybrids and their sexual hybrids *J. Agric. Food Chem.* 2007, *55*, 9089-9094.

[156] Fabroni, S; Ruberto, G; Rapisarda, P. Essential oil profiles of new *Citrus* hybrids, a tool for genetic citrus improvement. *J. Essent. Oil Res.* 2012, *24(2)*, 159-169.

[157] Parisi, A; Piattelli, M; Tringali, C; Magnano Di San Lio, G. Identification of the phytoxin mellein in culture fluids of *Phoma tracheiphila. Phytochemistry* 1993, *32*, 865-867.

[158] Salerno, M; Cutuli, G. Control of *Citrus* "malsecco" in Italy today. *Proc. Int. Soc. Citric.* 1977, 1001-1003.

[159] Ortuño, A; Reynaldo, I; Fuster, MD; Botìa, J; Puig, DG; Sabater, F; García-Lidón, A; Porras, I; Del Rio, JA. *Citrus* cultivars with high flavonoid contents in the fruits. *Sci. Hortic.* 1997, *68*, 231–236.

[160] Del Rio, JA; Fuster, MD; Gómez, P; Porras, I; García-Lidón, A; Ortuño, A. Citrus lemon: a source of flavonoids of pharmaceutical interest. *Food Chem.* 2004, *84*, 457–461.

[161] Belajovà, E; Suhaj, M. Determination of phenolic constituents in citrus juices: method of high performance liquid chromatography. *Food Chem.* 2004, *86*, 339–343.

[162] Caristi, C; Bellocco, E; Panzera, V; Toscano, G; Vadalà, R; Lezzi, U. Flavonoids detection by HPLC-DAD-MS-MS in lemon juices from Sicilian cultivars. *J. Agric. Food Chem.* 2003, *51*, 3528–3534.

[163] Nogata, Y; Sakamoto, K; Shiratsuchi, H; Ishii, T; Yano, M; Ohta, H. Flavonoid composition of fruit tissues of citrus species. *Biosci., Biotechnol., Biochem.* 2006, *70*, 178–192.

[164] Nogata, Y; Ohta, H; Ishii, T; Sekiya, K. Isolation of eriocitrin (eriodictyol 7-*O*-rutinoside) as an arachinodate lipoxygenase inhibitor from Lumie fruit (*Citrus lumia*) and its distribution in *Citrus* species. *J. Sci. Food Agric.* 2007, *87*, 82–89.

[165] Shaw, PE; Goodner, KL; Moshonas, MG; Hearn, CJ. Comparison of grapefruit hybrid fruit based on composition of volatile components. *Scientia Hortic.* 2001, *91*, 71–80.

[166] Moshonas, MG; Shaw, PE. Quantitation of volatile constituents in mandarin juices and its use for comparison with orange juices by multivariate analysis. *J. Agric. Food Chem.* 1997, *45*, 3968–3972.

[167] Rapisarda, P; Pannuzzo, P; Romano, G; Russo, G. Juice components of a new pigmented Citrus hybrid *Citrus sinensis* (L.) Osbeck × *Citrus clementina* Hort. ex Tan. *J. Agric. Food Chem.* 2003, *51*, 1611–1616.

[168] Sentandreu, E; Izquierdo, L; Sendra, JM. Differentiation of juices of clementine (*Citrus clementina*) clementine-hybrids and satsuma (*Citrus unshiu*) cultivars by statistical multivariate discriminant analysis of their flavanone 7-*O*-glycosides and fully methoxylated flavones content as determined by liquid chromatography. *Eur. Food Res. Techn.* 2007, *224*, 421–429.

INDEX

A

B

C

F

G

H

I

M

N

O

P

Q

R

S

T

U

V

W

X

Y

Z

Β